From test t tonnes

MW00893144

Commercial drug process development for life scientists

Dr CF Harrison

Introduction

Most people who pick up this book will have a science background of some sort. Perhaps you are finishing an undergrad or doctoral degree and want to start your first 'real' job. Possibly you are a postdoc, sick of hunting for grant money and looking to make the jump into industry. You have a scientific background – you know how to design and run experiments, data interpretation is not a problem, and it's been years since you last forgot the positive control.

But there is a vast step between the laboratory world and that of commercial manufacturing. It is a jump from plate readers and pipettes up to a world of thousand-litre fermenters and raw materials measured out by the tonne. A world where failure can cost millions and where anxious engineers sweat over the slightest variation in pH – and a world which most of us have never experienced before.

This book will cover the act of developing and validating a commercial process – from initial process design through to the lifecycle management. It will introduce you to the many steps which need to be taken, translate the strange language of manufacturing into real words, and explain just why everyone around you is freaking out when the process validation fails.

Sound useful? Then read on!

Table of Contents

Process development

So your company wants to make drugs? Well, there are many steps which need to be taken before you can actually sell at a commercial scale. Although we cover each of these steps in detail later in the book, this is the basic approach which every company will take:

- **Process Design**: You can't make something if you don't know how you're going to make it – and so the first step is to design your process. What raw materials will you need, how will they be combined, what sort of manufacturing operations will occur? How will you purify the final product? This step is all about deciding what you are going to do.
- **Risk Management**: You don't know everything, obviously. But do you know *enough*? Or are there large gaps in your knowledge which may lead to horrible failure later on? Risk management is a matter of looking at what is known and deciding where the risks lie – and then deciding what to do about it.
- **Process Control Strategy**: Your drug is going to be given to people – living, breathing people. This means that it needs to be the highest quality which you can realistically produce – no bacterial contamination, no birth defects, no endotoxin-mediated septic shock. Your control strategy is the process you will use to ensure the quality of the drug.
- **Qualifying equipment, facilities, and methods**: You need high-quality tools to do high-quality work. Qualification is the act of formally testing the various pieces of your process to show that they work correctly and consistently – and you'll be doing it for the factory, the equipment being used, even the analytical tests.
- **Process performance qualification**: This is the *test*, the test to see if your commercial-scale manufacturing process has what it takes. Can you make your drug with a sufficient level of quality? Consistently? If not, you will find out at this stage.
- **Continuing Process Verification**: The commercial process is all worked out, so now it's plain sailing – right? Not a chance. Testing never stops, and continuing process verification is your way of constantly checking for unwanted changes.

All of this work is done by the pharmaceutical manufacturer, and it's up to them to decide when they have enough data to justify commercial production. In practice, this will be a joint decision made by many people in different areas, based on many, many documented studies.

Once the company is confident, they will then need to convince the regulatory authorities (e.g. FDA, EMA, etc.) as well. This is done by providing information in a regulatory **dossier**. Authorities can agree, disagree, or (most likely) demand more information. In particular, the authorities will want to see that all stages of process design have been successfully performed – a successful process qualification run when you've no idea what you are doing is just as bad as a perfectly understood process that can't be performed.

The final goal of all of this effort is to be able to state, without too much hesitation, that you are extremely confident in your process. That you can consistently produce a 'good quality' product under commercial conditions. Being extremely confident is not easy, naturally, and so process validation in practice involves exhaustive amounts of testing, measuring, and writing reports. The following sections will cover this exhaustive testing in significantly-less-exhausting detail.

Examplifen: An example process

Developing a commercial process is, well, a pretty dry topic. So, in order to better show how everything works in the real world, we're going to demonstrate using an example process. Simplified, naturally, because we don't want this book to take up a few hundred pages, but enough to get the ideas across more clearly.

So, congratulations! You are now part of the process development team for our company's newest stomach-pain relief medication, the conveniently-named Examplifen. Examplifen is a small molecule drug (i.e. not a protein), which is designed to be taken orally (i.e. swallowed) and exerts its pharmaceutical effects directly on the stomach lining. The discovery and preclinical work has been looking good, the FDA has been optimistic in preliminary discussions, and we're going to assume for the purposes of our example that clinical trials will be completely successful without any problems (this never happens, of course, but it's an example). Thus the main problem now is to manufacture at commercial scale without difficulties. It's time... for commercial process development.

Process Design

The first step in making something is to know how to make it – and this is the problem that the Process Design step sets out to answer. Before we get into the details of it all, however, we're going to cover some basic terminology that you're going to need throughout your pharma career.

Manufacturing basics

Once a drug has exited the initial screening phases and becomes an actual pipeline molecule, the big question becomes 'how will we make it?' This can be a simple and well-established process or it can be highly complex with many variables to tweak. All production will need to be done under **Good Manufacturing Practice** (GMP), essentially a set of common requirements for making high-quality goods.

There are two parts to any pharmaceutical. The most important is the **active pharmaceutical ingredient** (API) – this is the part that actually does something (i.e. if there's no API, it's a placebo). The second part consists of the **excipients**, literally everything in the finished product that is not the API. Excipients are extremely important for making the drug work as it should and so are a major component of the development process.

Manufacturing can be thought of as a number of **process steps** linked together, each step indicating a single action in the manufacturing line such as mixing compounds, filter-sterilising, etc. The intermediate stages and materials will have a number of **quality attributes** (QAs) – properties or characteristics which affect the final product.

Scattered amongst the process steps will be tests to monitor the overall progression of the process and the quality of the intermediate material being produced. These attributes are affected by variations in **process parameters** (PPs), the term for measurable variables which occur during manufacturing. These process parameters are in turn monitored by **in-process controls** (IPCs), the measurements and values that will be taken as the process continues.

Confused? Here's a simple example: we mix the API with two secondary, non-active excipients in a buffer. The homogeneity of the final mixture would be a *quality attribute* – itself reliant on the *process parameter* of mixing time – while the act of measuring the mixing time would be an *in-process control*. The length of time that the mixing should occur is a target value, specification, or requirement. IPCs, PPs, and attributes are divided up according to their **criticality** (how much of an effect they will have on the final product's quality and safety), splitting them into the categories Critical, Key and Non-key

Multiple batches of drug product will be produced in a single **manufacturing campaign**, after which the line is often changed over to another product or strength. As each batch is finished it is packaged and labelled, then passed on to the quality group for batch release.

Development basics

Process design is the first step in the overall process validation cycle. The goal at this stage is to determine how the commercial manufacturing process will actually be performed, with an emphasis on developing a reliable and well-controlled process.

In practice this involves taking a lot of information from many different places and then bringing it together. This also means that you need a wide range of experts to interpret this information – chemists, engineers, microbiologists, statisticians, quality assurance, and manufacturing, to name a few. Thus you will be working with many different viewpoints and priorities during process validation activities, and need to accept that compromises will have to occur.

Although compromises can occur, one thing that cannot be compromised on is the need to document everything. *Everything*. There is a standard saying in the pharmaceutical world: "if it wasn't documented, it wasn't done". This applies from the very beginning. Every result should be recorded in a lab book and signed. Every study needs to be summarised in a signed and archived **report**. Even before you begin work you need a signed **protocol** setting out what exactly is to be done, how the data will be analysed, and what will be considered 'success'. Document everything!

Now onto the actual process design…

What do we already know?

The very first thing to do is to determine what is already known – what is known about the process and what is known about the product. This information is drawn from several typical sources:

- Development work
- Prior knowledge
- Process understanding
- Process modelling and simulation
- Scale-up

All of the information gathered in the course of these studies will be summarised in reports. The general approach used is to provide each separate experiment with its own protocol and report, while an overarching 'master' document or summary will link all of the important data and conclusions together. This is then combined with the general knowledge gained during development and the specialised knowledge of your local engineers to come up with a working commercial process plan.

The following sections will cover these sources in more detail.

Development work
The early development work will have given the company an initial idea as to how they will produce their drug of interest – which chemical reactions or microbial hosts will need to be used, how purification will be performed, etc. This initial plan will then need to be refined into one which matches both the equipment available and the complexities of the process. The initial studies will, however, have provided the company with an idea as to which parts of the process are more important than others – in particular which parameters need to be tightly controlled and which can be allowed to vary. This forms the basis of determining the in-process controls and process parameters which will be examined in later steps.

The initial research work is usually divided into two distinct steps, Discovery and Development, and Preclinical Research.

Discovery and Development
Discovery is the first step in the drug development lifecycle, and it can be thought of as the 'initial idea'. Maybe a new publication comes out showing an important protein in a certain disease, maybe the high-throughput screening finds a promising compound. Of the many thousands of potential candidates at this point, only a few will actually become a drug.

Development is the work you do after you've discovered a likely compound. There are a million different questions which need to be answered before you can move onwards. What is the mechanism of action? What benefits could it have as a pharmacologic? What would be the likely dosage? Are there other potential side effects? How should it be administered – orally? By injection? Rubbed into the skin as a topical cream? Are there certain side-groups on the molecule which could be adjusted for better performance? Or which might be metabolised in the body to create a toxic form? Large drug companies will have a number of processes set up to answer these questions as quickly as possible. The goal is to fail early (because failing early is *much* cheaper than failing later) or, if the early signs are good, quickly move on to preclinical development.

Preclinical research
Preclinical research is further split into *in vitro* and *in vivo* research arms. *In vitro* research is that done in the lab – think of the plate reader tests or tissue culture studies which help determine target binding rates. *In vivo* is animal-based research, studies which will focus on areas such as drug metabolism kinetics (ADME and the like) and toxicity. Kinetics and toxicity work will be used to decide just what drug dosages will be used when clinical trials begin (or even if clinical trials should be performed at all – many potential drugs will be stopped at this point).

There is no requirement to perform preclinical experiments under GMP conditions (which are focused on manufacturing) but the FDA usually requires that all preclinical work be done under GLP, or Good *Laboratory* Practice. GLP conditions are not as severe as GMP, but there is nonetheless a requirement that the work be done in a scientifically sensible and well-documented way, with qualified equipment, written protocols and SOPs, signed reports, and defined oversight from the quality group to ensure everything is kept in order.

Go/No-go decisions
All of the information gathered during these early phases will have two main uses. First, it will help to decide if the drug should go ahead for clinical trials (the 'big' go/no-go decision). This decision will be made by the very highest management levels as even small clinical trials can cost millions of dollars – a prospect which tends to make CEOs nervous, for some reason. Secondly, the information will help determine the

manufacturing process – by this stage you should have a good idea of how your drug will be made and what sort of quality parameters you need to be looking at. Although modifications will happen as you gain experience in the process, the first **clinical batches** need to be made under GMP conditions and with the best quality that you can manage.

Prior knowledge

Prior knowledge is a vital but often underrated factor in developing a commercial process. Almost every pharmaceutical company will have done something similar to the current project previously, (and if not, they will hire consultants who have). This means that at least one person will be able to point out the problematic parts of the process, the areas where things are likely to go wrong or the steps in which previous projects have had difficulties.

This does not always help, of course – every project goes wrong in its own special way. But it does tend to make the act of process design faster and tends to reduce the number of 'complete surprises' which occur.

Process understanding

Understanding the small-scale process is, funnily enough, the most important prerequisite for designing a reliable commercial-scale version. To achieve this, the company will run a number of small-scale experiments in which changes in all the possible variables are tested for their effect on the process and final product.

How do you know what variables should be tested? This is where the concept of **design of experiments** (DOE) comes into play. Design of experiments is a systematic method for determining which input variables affect the final output of the process – a somewhat complex way of saying that it helps determine cause and effect. DOE is based on the assumption that there are a number of *input factors*, both *controllable* (i.e. modifiable) and *uncontrollable*. The value of the controllable input factor is often referred to as the *level*. All of these factors feed into each process step and so affect the output of that process (known as the *responses* or *output measures*).

As would be familiar from your science studies, an experiment involves modifying certain controllable factors to see how the output measures change, then repeating the process to minimise error. DOE takes this a few steps further by examining multiple parallel variations in controllable factors.

For example, say the Examplifen mixing step has three potential controllable input factors: the stirring time, the vessel temperature, and the input flow rate. Each of these has two potential 'settings' or levels which they could occur at – we could stir for 5 or 20min, the vessel could be 25 or 28°C, the input flow rate could be 10 or 15 litres per minute. We want to determine which combination is best, as so devise a full-factorial DOE setup for these three factors as shown in the following table.

An example DOE setup where each of our three factors has two possible levels

Experiment	Factor 1: Stirring time	Factor 2: Vessel temperature	Factor 3: Input flow rate
1	5 min	25°C	10 L/min
2	20 min	25°C	10 L/min
3	5 min	28°C	10 L/min
4	20 min	28°C	10 L/min
5	5 min	25°C	15 L/min
6	20 min	25°C	15 L/min
7	5 min	28°C	15 L/min
8	20 min	28°C	15 L/min

Based on our knowledge of the process step we would devise several output measures, requirements such as 'homogeneity' and 'purity of the active ingredient'. These output measures would be measured for each experiment in turn and we would then use statistical tools to determine which factors correlate best with which output changes. This information provides a basis for further process optimisation and the incorporation of in-process controls.

DOE studies are very helpful in determining how different variables interact with each other, particularly when multiple factors can have cumulative effects on the final product. Because DOE studies with numerous variables (i.e. all of them in the real world) can be hugely complex and thus hugely expensive, all companies will perform a risk analysis before starting to determine which combinations will provide the most information and which can be left out. The results of these studies will be used to determine the parameter ranges used in the final process.

Process modelling and simulation

Another approach to determining variability uses either model systems or computer simulation – generally known as process simulation. Process simulation is usually first done for single steps within the process, this allows the approximations required to set up the simulation to be worked out on a relatively small scale. Simulations of each step can then be linked together for an overall 'process' model which can be used to plan out further 'real life' process changes.

A danger here is that compounding of the various small errors from each stage can provide completely wrong information at the end. This is particularly dangerous when dealing with other departments, as many people (especially management, but hopefully not you) will read more accuracy into the simulation results than they should. Thus regardless of what is modelled or simulated, you should *always* find out what the limitations of the simulation approach are. In particular, you should be able to understand the reliability and likelihood of the process predictions being made.

Scale-up

The information gathered during the earliest stages of development will be helpful in outlining the final manufacturing process, but will leave a number of gaps. The biggest of these is that the work to this point has been performed with small-scale equipment and volumes, work which simply doesn't relate to the large-scale commercial productions involving tens of thousands of litres of materials.

The increase in batch sizes is known as **scale-up**, essentially taking an existing process and using it to make a larger amount of drug. There are generally three main steps in the scaling process:

- **Laboratory scale**: The smallest scale and the scale at which all of the research and development will have been conducted. Think of beakers and stirrer bars, flasks full of bacterial cultures and maybe a small fermenter or two. Laboratory scale work is performed under 'Good Development Practice' conditions, which means you have certain requirements for recording results, calibrating equipment, etc. Certainly not as many as GMP, but it's not like an academic lab either.

- **Pilot scale**: Pilot scale is an intermediate scale, the point at which you are manufacturing using a process which is fully representative and able to simulate the full commercial scale. Essentially it is a miniature version of the eventual manufacturing process. Pilot scale production will usually allow extended testing or possibly clinical trials to be performed (provided the pilot plant is run under GMP conditions).

- **Commercial or manufacturing scale**: This is the scale at which you will be making drugs for the market. This is where the thousand-plus litre fermenters come in, alongside the industrial scale piping, high-speed filling lines and host of automation features. There are strict rules about production and monitoring, large amounts of raw materials being consumed, and batch sizes reaching tens of thousands to millions of doses. Commercial scale is the Real Thing, where even small changes in efficiency caused by your earlier decisions can end up making or losing hundreds of thousands of dollars' worth of product.

Moving between each of these scales will involve a **technology transfer**, in which a process is formally moved from one location and set of equipment to another (this occurs both during scale-up and when moving your commercial manufacturing from one site to another). Tech transfer is a complex process involving a lot of studies and test production (plus the associated pile of documentation), all of which should ensure that the method has been moved without a drop in overall quality or reliability.

Although it seems like a simple progression from small scale to large scale, it is actually quite complex – and a number of problems will pop up during this process. In general, the problems which arise during scale-up can be assigned to one underlying reason, and that is the fact that *things don't scale linearly*.

What does this mean? Suppose your initial development has identified that the Examplifen API must be made by mixing a certain amount of two chemicals together, producing a set amount of API and a little bit of heat. We could then, theoretically, mix a thousand times the amount of chemicals to produce a thousand times more API, and a thousand times more heat. Except that we can't, it doesn't work that way. Large tanks are more difficult to mix properly, the heat produced will have problems equilibrating (as we now have a lower surface-to-volume ratio), the fluid dynamics will be completely different. The larger process will require a different approach to the one used at small scale.

There are other variables beyond the scale which can have an effect on the process – this can be as obscure as different employees working in different ways or slightly different suppliers providing the same raw material. Some of these can be predicted, some will be a complete surprise. All of these factors mean that the initial scale-up process is difficult (and time consuming!) and limits the use of data developed during the development phase.

Some companies will have a **qualified scale-down model**, a small-scale version of the commercial process which has been shown to act in a similar way. Scale-down models are usually developed after the company has been working with their commercial process for a while and has the requisite knowledge to model it on a small scale. You will thus predominantly see them used to identify tweaks and improvements in existing processes which can be confirmed in later process qualification studies. Completely new processes will usually be too novel to develop such a model and so will need to rely on data developed during scale-up.

Examplifen: Process design
We now return to our Examplifen example, with the aim of developing a manufacturing process. First we will take the prior knowledge which we have gained to determine what the final criteria of the drug product should be. From that point we will develop a manufacturing process. Later on in the book, we will use the knowledge here to develop our quality control strategy.

Knowledge from development
During the course of developing Examplifen, our lab people have come up with a number of parameters which maximise both product stability and efficacy. Naturally, these are our goals for the commercial product as well. They are as follows:

   a) Preclinical studies covering drug pharmacokinetics have shown that the ideal concentration of Examplifen for quick acting pain relief is 100μM. Thus the final drug product should have an Examplifen concentration at or around this level.

b) Stability studies using different pH values have shown that the API (Examplifen) is most stable at a pH of 3.2. The final product should thus be kept buffered around that pH in order to reach the maximum possible shelf-life.

c) Our buffering system is important for long-term stability and clinical efficacy. Early development work tested a number of buffer systems, and showed that the citrate-phosphate buffer system worked best when composed to make a pH of 3.2. Thus the final concentrations of excipient should be 0.049 M Na_2HPO_4 and 0.0753 M citrate.

Knowledge from previous experience

Our company has previously developed a number of medications and so we are able to make several suggestions for the final product based on this experience.

a) Survey of patients and readers of this book have shown that people prefer calculations which use round numbers. As such it is preferable that each bottle will be a single-use container and will hold exactly one millilitre of the Examplifen drug product

b) The final drug product is an orally available medicine, which means that the sterility of the product isn't as important as, say, an injectable vaccine (stomachs are full of bacteria, after all). However, it is still important that the drug be comparatively free of bacterial contamination. Our scientists and regulatory affairs groups say that there should be no more than 10 bacteria per millilitre of final product, which can be easily achieved via autoclaving of the filled and sealed bottles.

c) Large chunks of undissolved components both look bad and can be a health hazard. To avoid this, the compounded solution should be filtered before it is filled into the bottles.

d) As the concentrations of excipient and API are so different, our previous experience has shown that it is better to create two separate solutions and then combine them later in a **compounding** (final mixing) step.

Release specifications

Based on our development work and previous knowledge, we thus have a good idea what our final product should look like. This is expressed as a number of target values known as **release criteria**, the values which need to be reached for the product to be *released* to the market.

Release Parameter	Target Value
Concentration of API (Examplifen)	100 μM
Volume filled	1 ml
pH	3.2
Concentration of excipient (citrate)	0.049 M
Concentration of excipient (phosphate)	0.075 M
Bioburden	≤ 10 CFU/ml

This table shows the target release criteria for our drug. It is essentially a restatement of the previous information, but it now acts as a *goal* for our commercial process – it is what we want to achieve before we can release our drug to the market. It is also stated somewhat more formally, in particular the allowable bacterial contamination is now restated in the universally-used terminology of **bioburden** in **colony forming units (CFU)** per millilitre.

No process is perfect and no measurement will ever line up exactly with these target values. Thus the final release criteria will instead be expressed as ranges within which the values need to fall. These are known as **acceptance criteria** – if a value is outside the range, it's no longer acceptable. These acceptable ranges come from the earlier development work, which will tell you how much you can vary from your target value before problems start occurring. Parameters which are less important or have less of an effect on product quality can have a larger range than those with a major effect on product quality.

The release specifications now look as follows:

Release Parameter	Target Value	Acceptance Range
Concentration of API (Examplifen)	100 μM	90 - 110 μM
Volume filled	1 ml	9.5 - 10.5 μl
pH	3.2	3.1 - 3.3
Concentration of excipient (citrate)	0.049 M	0.044 - 0.054 M
Concentration of excipient (phosphate)	0.075 M	0.070 - 0.080 M
Bioburden	$\leq$ 10 CFU/ml	$\leq$ 10 CFU/ml

The Examplifen manufacturing process

Based on our knowledge and the requirements listed above, we can put together a preliminary manufacturing process. There are several points which need to be achieved:

- We should make separate excipient and API solutions prior to compounding.
- The pH needs to be adjusted to the correct range for long-term stability.
- The compounded solution needs to be filtered before the bottles are filled.
- The filled and sealed bottles need to be autoclaved.

This leads to our overall manufacturing process, which is shown in the process flowchart here. Actual manufacturing processes are normally a bit more complex than this, thus you end up with much more complex flowcharts – nonetheless the approach remains the same.

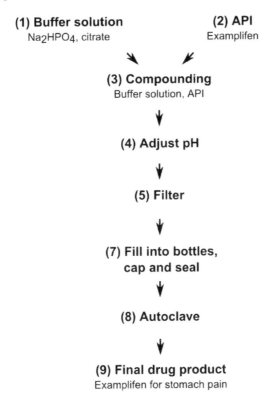

The initial manufacturing flowchart for Examplifen. The fairly simple description we had before still turns into a large number of process steps.

It's immediately apparent that even a simple process such as the one we are dealing with here still contains a number of steps to keep track of. 'Real' processes tend to contain many such steps, and this in turn is why process validation is such a tricky task.

Risk management: Filling the knowledge gaps

If there is one word which you are guaranteed to hear during your pharmaceutical career, it is "risk". Risk assessments, risk managements, risk reduction – risk will come up at every stage of the product lifecycle.

So then, what exactly *is* risk? Risk is a way of balancing the multitude of different things which *may* go wrong, known as **hazards**. To allow this balancing, risk is considered to be a combination of two different factors – the probability of *occurrence* (how likely it is that something will happen) and *severity* (how bad it will be when it happens). This allows you to compare common-but-harmless events to rare-but-deadly ones and so rationally decide where you should focus your efforts.

Risk is notoriously hard to quantify. This is partly because there are so many different things which could go wrong, but also because everyone sees the individual components in a different light. Your local doctor will disagree with QA as to the severity of the risks which have been identified, upper management will see different priorities for risk reduction than production engineers. There are also a number of different focuses – risks can be a problem for the company, for employees, for the patients, for the environment, and much more.

Health authorities cut through all of this confusion by simply stating that the *safety of the patient* is the most important factor – and thus that managing risks which may affect drug quality should be your number-one goal. Thus although you may have a number of risk management programs going on at once, a safe drug is the top priority – even in cases where the added costs will hurt the company.

This all sounds very complicated but it can be boiled down to two simple principles:

(1) Use scientific knowledge to evaluate your drug quality risk, with patient safety as the ultimate goal.

(2) The amount of effort you put into managing the risk should be proportional to the severity of the risk.

Easy, right? Let's go into a bit more detail.

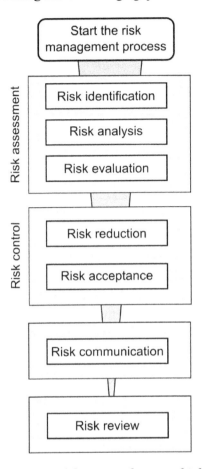

An overview of the general steps which are involved in performing a risk assessment.

Who is involved?

To evaluate risks you need experts, and as risks come in many different forms these experts need to be drawn from many different fields. This means that the risk management process will invariably be performed by interdisciplinary teams, involving experts from engineering, quality, regulatory, production, statistics, and many more. Mixed in amongst these will be several people with experience in risk management to help guide the process.

Although risk management is a very important thing for companies to do, it isn't something that requires full-time staff. Thus the team will usually be ad-hoc, they are brought together to perform the assessment and then return to their normal jobs when the assessment is complete.

Before you start

A risk assessment is like any other project, and thus the standard project management rules apply. Before you even think about risks, you should decide the following:

(1) What are you actually assessing?

What problem or risk-related question are you trying to solve or answer? What assumptions are being made in order to answer these questions? You need to know this in order to be sure that your risk assessment will actually look at the right things.

(2) What do you already know?

What background information do you already have which may relate to potential hazards or risks? Has data been generated from other studies? Most importantly, is the data or information relevant to your problem? A clear definition from Step 1 will prevent you from being bogged down in useless information.

(3) What do you have to work with?

Who exactly will be on your team? Who will be the leader/co-ordinator, responsible for getting the whole thing across the finish line? Who do you know that is *not* on the team but who can be tapped for specific advice if necessary?

(4) What do you have to do?

What will you actually have to produce at the end of this entire process? For example, are you expected to complete a risk assessment, provide a risk management plan, compile a list of suggestions for improving the process? How long do you have to get everything done? Are you allowed to make decisions and changes based on the assessment, or do you need to work with management?

Clarity at this step will save you a *vast* amount of trouble later on, so make sure that you can answer all of these questions before you move on to the actual risk assessment.

Step one: Risk assessment

The basic approach in risk assessment is to first determine what *could* go wrong (the potential hazards), and then to determine what the risk associated with those possibilities are. Quality risk assessments, which focus on hazards to the drug product quality, follow exactly the same approach. As mentioned before, you need to have a well-defined problem or question you want to answer – this forms the basis of all the following work.

Risk identification

Risk identification is the formal terminology for the process of saying "what could go wrong?" A lot of things could go wrong, many of which you won't think of at first. Thus a successful risk identification process will involve looking at historical data, asking experts for their opinions, checking development studies, hunting down theoretical proposals, and much more. The aim is to *systematically* map out the potential hazards in your process and identify possible consequences of their occurrence.

Risk analysis

Now that you've put your long list of possible hazards together, it's time to decide just how much of a risk they are. There are several ways of doing this (which we describe in the appendix), but most methods will use a combination of three factors: the likelihood that a hazard will occur; the consequences of the hazard occurring; and the chances that you will detect the hazard.

Risk evaluation

This is the point where you begin to compare all of the risks you've analysed against your predetermined evaluation criteria. Basically you are looking to decide which ones you will consider 'major' risks and which ones are 'minor', which need to be dealt with and which can be accepted. This decision is often based on the resources you have available – with infinite money you could solve all of the problems, but in the real world you need to prioritise.

If you want to do an accurate risk assessment (and, let's be honest, you should), then you need to be working from accurate information. More importantly, you need to know where the information *isn't* accurate – where you have made assumptions to cover gaps or uncertainty. Uncertainty is not a terrible thing, it's just a sign that you simply don't know everything about your process and the science behind it. The question then becomes whether that uncertainty is something you can live with, or whether further studies need to be planned to fill in the gap.

Risk itself can be given in a number of ways. The simplest is a qualitative description – high risk, low risk, boringly negligible risk. More usefully, you can have quantitative systems which base their ratings off calculable numbers such as probabilities. The most common approach uses **relative risk** based on combinations of the underlying assessments (if you aren't familiar with relative and absolute risk, we quickly explain the two on page 28).

Step two: Risk Control
Gone through the process and assessed all the risks? Great! Now you need to decide how you will actually control all of those risks and bring them down to a reasonable level. This is the idea of **risk control**.

The important thing about this step is that the efforts taken to control a risk should be proportional to the severity of that risk. There is no point spending millions of dollars on preventing a hazard which rarely happens and has basically no effects when it does – that money should instead be spent on preventing nastier risks which can lead to things like patient deaths.

This leads directly to the two basic approaches which can be taken when confronted with a risk. You can choose to *reduce* it or to *accept* it – and both of these have their ideal time and place.

Risk reduction
Risk reduction is all about, well, reducing risk. This may be through extra testing or tighter limits during the process, higher criticality levels for IPCs, even complete changes in your manufacturing approach. There are lots of different ways in which risk can be reduced – the trick is to find the best one. This is where the knowledge which you gathered at the start comes in, as it provides the basis to scientifically choose the best reduction approach.

Risk reduction tends to suffer from diminishing returns – it's comparatively easy to reduce risk by, say, 50%, but the next 25% will be much harder. Because of this trend most programs will try to get down to an *acceptable level*. What is the acceptable level? Well, that depends on how severe the risk is, how many other risks you have to deal with, and what sort of resources you can throw at the problem. As in every part of pharma, it's about balancing different requirements.

Another thing to watch for at this stage is newly added hazards. Adding a risk reduction method may lead to other problems or risks (e.g. you take more samples to monitor your purity better, but now you have a higher chance of product contamination). It's often worth redoing the risk assessment after you've put all your ideas in place to ensure that no new problem has popped up.

Risk acceptance
Risk acceptance is the decision to say "yes, it may happen, but we are ok with that." This is perfectly fine if you can back it up with scientific knowledge – statements such as "it may happen, but only once per thousand years" or "it may happen, but it won't really be a problem", or even "it may happen, but we'll spot it straight away." These are clear risk acceptance decisions made on the basis of prior knowledge, they will be stated in the risk management plan and agreed to by various experts. The problem comes when risks are accepted because you don't know enough – you think that the hazard is less likely or less dangerous than it actually is. Sometimes you may not even realise a hazard exists – the "unknown unknowns" which cause the biggest problems.

Risk acceptance also comes into play once the risk reduction processes have been implemented. You are basically never going to completely remove risk from your drug manufacturing process – something can always go wrong. The aim is to reduce risk to the point mentioned above, where you are 'ok' with the chance that the hazard will occur.

Step three: Risk communication
There's no point doing all of this work if no-one ever finds out about it. Risk communication is thus about taking all of your findings and passing it on to people who can actually *do* something about it. This usually means management, but you'll also want to involve production, quality assurance, regulatory affairs, etc. – basically anyone who can help push changes through.

In the formal sense a risk communication is done by passing on the nicely written and thoroughly signed risk assessment report, complete with conclusions and recommendations. Though nice, the majority of your information will be passed on in various meetings and teleconferences where everyone scrambles around to block the gaps which has just been found. Expect to see a number of **CAPAs** (Corrective and preventative actions) being assigned to various experts, usually requiring SOP and process updates to solve problems which were noted.

Step four: Risk review
Risk assessment is complete, report done and signed, management informed. Top show! Time to knock off and have a beer or two. And then, once you get back to work tomorrow, then it's time to get back into the **risk review** part of things.

Risk assessment is an ongoing process, because every bit of experience you gain in your process will give you more information on potential hazards. Perhaps one step is less risky than you thought, or an IPC is more likely to give a false reading. This kind of information helps you to refine and optimise your risk management approach over time.

In practice you will do periodic reviews of the risk assessment every year or two, more often in the case of particularly risky process steps. This will incorporate all of the new information you've gathered since the last time, and will often bring up new ideas and updates which will provide better risk control.

Relative risk versus absolute risk
There is an important distinction between relative risk and absolute risk, and it's one you need to be aware of during this whole process. Absolute risk is the chance that something will happen – if you have a manufacturing process and one in every ten of your batches explodes for no apparent reason, then you have an absolute risk of 10% (you can also state this as 1 in 10 or a probability of 0.1). Relative risk is the chance that something will happen – *in comparison* to another situation. Thus if we take our above process and begin using a new raw material, only to discover that we now have one in every five batches exploding. This new raw material gives us an absolute risk of 20% but a 2-fold increase in relative risk – it is now twice as risky as before. Always be aware of whether you are working with relative or absolute risk measures when dealing with risk assessments.

Risk Assessment Methods

There are a number of tools which have been put together to help out with risk management, each of which have different focuses and different advantages. Information on the most common of these methods is provided in the Appendix of this book.

Establishing a process control strategy

There are a number of rules and regulations surrounding the manufacture of pharmaceuticals. Almost all of these focus on ensuring that the final drug is of sufficient *quality* – that the patient can be assured that the drug will work as expected every time and without any unexpected side effects. In other words, the drug should be fit for its intended use.

Because drug quality is so important, most manufacturers use an integrated approach to controlling the quality of all aspects of their process. This means that the product and manufacturing line are not just reliant on a series of measurements to keep the process running as it should. Instead, they are designed in such a way that all of those quality, safety, and efficacy factors are built in – the process should run the right way every time.

So how do you actually achieve this? Once you have learned as much as possible about the process itself, then it is time to decide on the **control strategy**. There are two main approaches which are taken when putting together the control strategy, often referred to as Traditional and Enhanced by regulatory authorities. Ideally both of these approaches will be incorporated into the final process.

- *Traditional* approaches set operating limits for various process parameters, then perform a series of validation runs to prove that these limits can be met. A number of tests are performed in standard manufacturing to make sure that the variables stay within the limits at all times.
- *Enhanced* approaches require a much better understanding of the whole process, and use this understanding to develop a **design space** in which the various parameters should lie. A design space is best visualised as a combination of different parameters – your measurements are no longer staying within a one-dimensional range but rather a multi-dimensional area.

Each process can be broken down into basic steps known as **unit operations,** such as mixing or filtration. Each of these unit operations has associated checks and balances, generally known as **process controls**. Process controls can cover materials and equipment, and the general assumption of regulatory authorities is that all of the important process steps will have some sort of process control associated with them. They are considered particularly important in places where measurement is difficult to do (bacterial contamination, for example, is difficult to spot), or where the intermediate being checked is relatively heterogeneous and hard to characterise. These important controls will require tight limits to be clearly stated in the production plan.

The actual act of developing a manufacturing process and associate control strategy will differ from company to company and drug to drug. There are some minimum checkpoints which need to be met – you'll have to identify the critical quality attributes (CQAs), the process will need to be developed to work with the CQAs, and the control strategy will need to be put together. Usually you will also need to determine how the properties of the raw materials and process parameters fit together to control the quality of the final product.

Critical quality attributes (CQAs)

So what exactly is a CQA? Simply put, it's any type of property or characteristic that needs to be within a certain range – if you go outside the range, the quality of your final product suffers. No-one will know all of the CQAs when they start developing the manufacturing process, generally you will start with a list of 'potential' CQAs which will be modified as experience is gained.

Actually deciding which of your quality attributes should be considered 'critical' is a difficult task, particularly when complex processes are involved (i.e. all biological ones). This is usually done via the trusty risk assessment, where experts will determine where in the process the greatest quality risk lies and thus where they should focus the most attention.

There are several 'typical' quality attributes which will pop up in every drug product and every manufacturing process. These tend to be considered crucial by regulatory authorities and so are always closely watched. They consist of *identity* (are you making what you think you are making?), *quality* (is it well made and does it hold up over time?), *purity* (is it just the drug, with no unwanted contaminants or breakdown products?) and *potency* (does it have the effect that it should?). It's important to remember that there is a difference between *strength* and *potency* – strength is the amount of drug in the table (e.g. X mg aspirin), potency is the effect it has on the patient (e.g. lowers blood pressure by X).

Of these, one of the most important CQAs in any process (and one which will cause many frustrating moments) is that of *purity*. **Impurities** can be organic (carbon-based) or inorganic (such as heavy metals), they can also be **residual solvents** left over after manufacturing steps. Biological products can have impurities from the cell line (such as Host Cell Proteins or DNA), from the fermentation process (media components), or even, ironically enough, from the act of purification itself (think of buffer components or compounds leaching from the purification columns). Impurities will also overlap with **contaminants**, things which shouldn't be in the product at all. Typical contaminants will be bacterial or viral contamination.

Linking material attributes and QCAs

The quality of the starting material affects the quality of the final product – something which has been obvious to anyone who has tried to cook on a student budget. Thus as you'd expect, if you want to have a high-quality final drug (which you do) then you should control the quality of the raw materials.

More specifically, this usually focuses on controlling the levels of impurities which are present or which may be introduced during manufacture. How much effort goes into this is very dependent on how detectable it is and how hard it is to remove. A minor residual solvent which is completely cleared in the first process step is much less important than the host cell protein which keep co-eluting with your biological therapeutic. Similarly contaminants such as bacteria or viruses, both very difficult to detect in small samples, need more attention that something obvious such as heavy metal contamination.

Design space

Design space is a way of combining a number of process parameters and limits to create an 'area' in which you can work without worrying about quality problems.

What does this actually mean? Suppose the purity of a process step output is dependent on two factors, the pH and the concentration of sodium ions. Traditionally, you would simply set two acceptable ranges for the two factors – any batch which went outside the limits would then trigger a deviation and would probably lead to it being rejected. However, further studies showed that the two parameters were linked, a high pH would lead to a pure compound as long as the sodium concentration was low enough. This now leads to a 'design space' in which the pH and Sodium concentrations are plotted, combined values within this region mean that you are ok to continue.

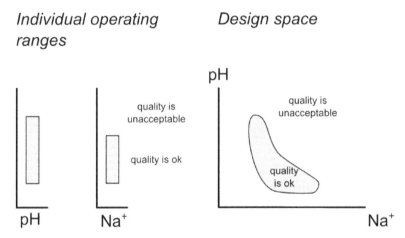

In contrast to individual operating ranges, a design space allows the combination of measured values such that the overall quality of the product is assured.

Although this requires a lot of extra work to fully understand the process, having a design space approach makes manufacturing much easier for the company. Whereas before a batch which was outside the limits would need to be rejected, the new system allows it to be salvaged so long as the matching parameters are ok. Even more usefully, parameters from different steps can be linked together into one design space – thus a low pH value in Step X may be countered by increasing the Sodium concentration in Step X+4 and thus providing maximum control over quality.

Design spaces are most useful when comprised of several different parameters. The following figure shows relation of several different parameters on the time taken and temperature during a process step. The acceptable region, shaded in grey, is slightly different for each parameter – be it bioburden, purity, or adventitious viral contamination. The *design space*, in which the parameters can be modified without any problems in product quality, can be found at the intersection of these regions.

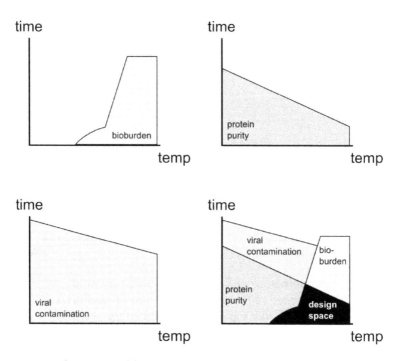

The acceptable range for several different criteria are plotted as a function of time and temperature during a process step (shaded). The appropriate design space can be taken as the intersection of these different functions, as seen above.

In general, materials and process steps which occur towards the start of the manufacturing process will be less risky than those which come later on. Impurities which are brought in at the start have more chances to be removed through purification steps than those which occur directly at the end. Similarly the most important physical properties of the final drug are set in the last manufacturing stages, and so problems here lead to a much higher risk than those occurring at the beginning.

Control strategy

A **control strategy** is a general term for all of the monitoring and testing which you do to ensure the quality of your final product. A manufacturing process without a control strategy is the equivalent of two guys in a basement making meth – you may end up with something but you will never get regulatory approval.

Being a broad term, a control strategy will often cover a number of more specific controls such as:

- Material attribute controls: This covers all of the testing which will be done on raw materials, packaging materials, reagents, etc. This testing is normally done as *incoming goods testing*, testing which is performed when the material is delivered to the company.
- In-process controls: Usually shortened to IPCs, this is testing which happens *during* the manufacturing process. There are usually several IPCs per process step and they serve to keep the entire production run on track.
- Final product controls: This is also known as **release testing**, it is the series of tests and required specifications which need to be passed in order for the product to be *released* to the next stage.

Developing a control strategy is usually done through a combination of traditional and enhanced approaches. As mentioned previously, the traditional approach involves setting fairly narrow ranges for each process parameter based on previous observations, with a lot of importance being placed on the final product controls. Enhanced approaches build on improved process knowledge to use a more flexible approach – by knowing how variables and parameters interact with each other it is possible to modify the process as you go to cope with changes in material attributes and IPCs.

Regardless of how it is done, a control strategy should be able to keep all of the CQAs within their required parameter ranges. This can be done as part of release testing, through upstream testing such as IPCs, or at both stages. Variable or difficult to detect CQAs will be checked at multiple points during the process as well as during release (e.g. bioburden will usually pop up on the final specification sheet as well as throughout the manufacturing process). Important yet reliable CQAs can be checked at release – the concentration of active ingredient in the final drug product is important but unlikely to vary much during manufacture. Finally CQAs which are only relevant for part of the process may be checked using IPCs and left off the specification sheet entirely.

Risk assessments

At heart, a process control strategy is simply a way to manage risk – things may still go wrong, but your system should be able to minimise the risk that a bad batch will reach the patients. Thus an important part of designing the process controls is the risk assessment, or rather risk assessments – you'll be doing quite a lot of them soon! The general approach for risk assessment was covered in more detail in the previous sections. In general, however, a risk assessment at this stage will look at a number of areas:

- Raw materials: How variable are the raw materials being used? What impact will that have on the process? How is this variability controlled?
- Equipment capability: How good is the equipment which you are going to use to make these commercial batches? Is it good enough to get the job done? Consistently?
- The process itself: Has our earlier work shown the process to be reliable or is it difficult to work with? Is the scale-up process likely to have an effect? Do we have past experience on similar things to help get us past problems?

All of these factors will be taken into account when deciding how the process will be set up and how much extra effort will need to be done for process qualification – a riskier process will need more tests.

Process analytical technology

A somewhat fancier method of controlling a manufacturing process is known as **process analytical technology** (PAT). The underlying theory here is that the typical manufacturing approach (in which batches are produced one at a time and samples collected during production are tested afterwards) is reliable but outdated. PAT, by contrast, follows the philosophy that quality cannot be implemented by testing alone, but instead should be designed into the process from the start.

This is in practice achieved by a number of different approaches, most notably by understanding your process really, really well. However, it also involves the implementation of 'real time' process controls – in which you take your measurements during the production process and use that information to modify the process as you go (think of slight modifications in mixing time based on real-time measures of homogeneity and the like).

Regulatory authorities, the US Food and Drug Administration (FDA) in particular, are trying to push drug manufacturers towards process analytical technology methods. From a regulatory point of view they are preferred due to the better level of control provided over the process, given the semi-real-time feedback and adjustment. Similarly from a company point of view PAT systems are an advantage, the ability to catch problems as they occur means that batch reject rates can be significantly reduced. They do come with disadvantages, however, including high implementation costs and a corresponding increase in complexity for the process validation stages.

Designing a quality control strategy for Examplifen

In process controls and process parameters are the backbone of your quality control system. They provide an insight into how everything is progressing while the manufacturing is still going on (as opposed to release testing, where you will only find out at the end). Every manufacturing process will have a number of IPCs to cover each step, with a particular focus on the most risky areas.

To avoid making this too long, we will focus our example on a small subset of the Examplifen IPCs.

Amount of excipients and active pharmaceutical ingredient:

Through an amazingly convenient coincidence, the commercial batch size of Examplifen is planned to be about 100L, while the molecular weight of the API is 100 Daltons. As the concentration of API should be 100 µM, every batch should thus have 10 mmol of Examplifen, or 1 g. As you would expect, this means that every batch should have 1 g of Examplifen added during the initial compounding stage. The normal approach would be weigh this out and dilute into a separate solution, which is then combined with the other excipients during final compounding.

Similar logic applies for the excipient buffer components. The final concentrations of excipient should be 0.049 M Na_2HPO_4 and 0.0753 M citrate, thus a 100L batch would require 4.9 moles of Na_2HPO_4 and 7.53 moles citrate – or 695.6 g and 941.4 g. These will be diluted into a second vessel and completely mixed with the API in the compounding vessel.

Nonetheless, this provides us with several acceptance criteria which we should work within:

Step Number	Process parameter/ In-process control	Acceptance Criteria
2	Amount of Examplifen	0.99 – 1.01 g
1	Amount of Na_2HPO_4	695 – 696 g
1	Amount of citrate	941 – 942 g

pH:
The final pH should be 3.2, and there is only one step where we can adjust the pH (Step 4). This means that, logically, the target value of the pH adjustment should be 3.2 and the associated IPC after adjustment should be 3.2. Or rather, as we're using ranges, pH 3.1-3.3.

Step Number	Process parameter/ In-process control	Acceptance Criteria
4	pH of adjusted solution	3.1 – 3.3

Filter integrity
We have a filtration step in order to make sure any chunky bits in the compounded solution are removed prior to bottle filling. As part of this, we need to be sure that our filter has actually worked and hasn't suddenly developed tears or holes. The typical way to check this is a pre- and post-filtration **filter integrity test** – there are several methods here but the most common is the *bubble point test*, in which the pressure required to force bubbles out of a filter is measured. The actual pressure required will be different for each type of filter, our filters here should be around 1000 mBar.

Obviously our filter should still be intact, so this gives us two IPCs, one which we check before we filter everything (in case there's a problem) and after we filter (in case something tore during the process).

In the end, we have two potential in-process controls relating to filtration:

Step Number	Process parameter/ In-process control	Acceptance Criteria
5	Filter integrity test prior to filtration	950-1050 mBar
5	Filter integrity test after filtration	950-1050 mBar

Bioburden:

Bioburden is one of the most important parameters in any medication. Growth of even benign bacteria can affect the final concentrations of excipients and APIs, pathological bacteria can lead to patient illness or death. The bioburden of the final drug product, therefore, is a very important factor.

What level of bioburden is allowable depends entirely on the use of the final drug. An injectable product will be directly inserted into the body, it therefore needs to be completely sterile. The conditions in the stomach are perfect for destroying bacteria, thus medication which is taken orally has lower requirements for bioburden.

We have determined that we want less than 10 CFU/ml in the final bottle (our release test). Our autoclave process is likely to easily meet this, as long as the sterilisation is performed correctly and that the initial material is relatively bacteria-free. This means we have several IPCs relating to the bioburden process.

- The bioburden of our compounded solution should be low enough that later autoclaving can take care of it. A typical sterilisation process is at least a 6-log reduction, i.e. 1 out of every million bacteria will survive. Therefore a compounded solution bioburden of 1000 CFU/ml will easily be low enough to have a highly sterile final product.
- The autoclave conditions should be met for each run. More than 20 minutes at 122°C is a fairly typical sterilisation condition

which should suffice for this process, so our acceptance criteria is to meet this requirement.

We can summarise the acceptance criteria as follows:

Step Number	Process parameter/ In-process control	Acceptance Criteria
3	Bioburden of compounded solution	≤ 1000 CFU/ml
8	Autoclave conditions	≥ 20 minutes at 122 °C

The quality control strategy

In the end we have a list of process parameters and in-process controls which make up part of our quality control strategy, as listed below. These will be tested for every commercial batch which we produce, any results that fall outside the required acceptance criteria will trigger a deviation investigation and very often force the batch to be rejected.

Step Number	Process parameter/ In-process control	Acceptance Criteria
1	Amount of Na_2HPO_4	695 – 696 g
1	Amount of citrate	941 – 942 g
2	Amount of Examplifen	0.99 – 1.01 g
3	Bioburden of compounded solution	≤ 1000 CFU/ml
4	pH of adjusted solution	3.1 – 3.3
5	Filter integrity test prior to filtration	950-1050 mBar
5	Filter integrity test after filtration	950-1050 mBar
8	Autoclave conditions	≥ 20 minutes at 122 °C

Pre-Process Qualification

Process qualification is the act of testing the process design which was put together in Stage 1 and seeing if it is up to scratch. All of the work at this stage is done under GMP conditions, which means that the final product can actually be sold on the market if the regulatory authorities approve your process in time. Having said that, there are many studies which need to be performed before the process qualification can actually occur – think of them as the foundations upon which the final check is built.

'Pre-process qualification steps' include several very important studies:

- Qualification of facilities
- Qualification of equipment
- Qualification of the analytical methods

Qualification of facilities

Good manufacturing starts with a good building, and thus the first thing you should be sure of is the factory itself. You should remember that in the real world there will often be a significant overlap between facility and equipment qualification – particularly in pharma where clean room facilities are required. Here, however, we will look at them separately.

Regulatory authorities have a number of general requirements for buildings which fall under the scope of basic GMP, and this has been boiled down into three steps known as Design Qualification (DQ), Installation Qualification (IQ) and Operational Qualification (OQ).

Design Qualification (DQ)

A building starts with a plan, and in the pharmaceutical world this is a very detailed plan indeed. The **specification** defines all of the requirements for the building and the processes to be held inside, and it includes a number of details such as:

- *The standards and regulations which will be followed*: There are usually a number of regulations which apply to any one facility and so this can be quite a confusing list.
- *Facility description*: What will be done in the facility? What are the process steps? What requirements for inputs are there?

- *Mechanical components and materials required*: The process will use machines (hopefully!) and so you should define which machines and equipment will be included. What do those machines need to be made of (e.g. stainless steel for mixing vessels)? How much power will they draw and how much heat will need to be removed?
- *Dimensions*: This one is quite self-explanatory – how big should the rooms be? The facility? The pipes?
- *Controls and process monitoring*: You need to be able to watch over the manufacturing process as it runs, and thus a specification will include such details as the sensors and display outputs which will be needed.
- *Energy and utility supply*: Need electricity? Of course you do. But what about hot water, or constant high pressure air, or even a nitrogen flow to prevent oxidation? All of these need to be specified.
- *Installation*: A specification will also double as an agreement between the pharmaceutical company and the manufacturer, and so it will include extra information on the delivery/handover process. This includes the definition of the Site Acceptance Test, the final verification from the company that everything is ok and that they are willing to take over responsibility for the site.

Pharmaceutical facilities have far more stringent requirements than other factories do, and regulatory authorities such as the FDA tend to require a lot of extra work. In particular they look for cleanliness, organisation, and flow.

Pharmaceuticals have high requirements for *cleanliness*, unsurprisingly enough, and this also extends to the production area. Cleaning is particularly important in zones which are considered to be clean or aseptic. These areas will have hard, easily cleanable surfaces, temperature and humidity controls, and often HEPA-filtered air. A cleanroom is an isolated zone which is very closely monitored for airborne particles, microbial contamination, airflow, and much more. They use specially filtered air and controlled airflow to ensure low particle levels and extra cleanliness for critical equipment. Keeping everything clean is vitally important, and so production zones will have regular **environmental monitoring**. Environmental monitoring involves checking for airborne particles, checking microbial contamination on surfaces and in the air, even checking the microbes on the workers themselves.

The building needs to have sufficient *organisation* that there is no mix-up or contamination of materials. Contamination is a major worry for regulatory authorities and so smart pharmaceutical manufacturers make efforts to prevent all possible contamination events. This sounds complex but really comes down to a few simple rules – label everything really well, store different materials in clearly defined places, and have clearly defined standard operating procedures for everything. Everything else stems from these basic rules.

The *flow* of goods, air, and material needs to be designed so as to prevent contamination. Raw materials will enter the production area, run through a variety of production steps as they are used to create the pharmaceuticals, and the final product will then exit the production zone. Workers will start their shifts, move around to keep everything running and under control, and eventually go home. This movement throughout the production area needs to be planned out such that the potential for cross-contamination or accidents is minimised.

Airflow is also an important part of controlling contamination. A system will usually be set up such that the cleanest zone is the one with the highest air pressure – this prevents contaminants from drifting in from less-clean areas and so protects the clean zone. This is known as a *pressure gradient* or *differential pressure*, and it should not be confused with a *laminar flow* approach. In laminar flow, the air supply is designed to create a uniform flow in a single direction which can push particles along with it. This will ensure that any particles or dirt which do enter the clean zone will be quickly blown outwards towards the dirtier zones.

Once you've gotten all your specifications in place (and found someone to create the facility) then it is time to move on to installation qualification.

Installation qualification (IQ)
Installation qualification is the act of showing that the facility and components have been put together correctly. Basically, if you intended to order something, then it should have been installed – and the

installation qualification will be the act of proving that. It is an extremely thorough process which will cover all aspects of the building in question, but the major discussions will revolve around a few areas:

- *Technical documentation*: Good documentation makes life infinitely easier – just ask anyone who has tried to put Ikea furniture together without the instructions. As in Expedit so in pharmaceutical facilities: the technical documentation is a must for effective handover. Building plans, detailed operating instructions, maintenance manuals. All of it needs to be present.
- *Standard operating procedures*: The SOPs involved in performing typical maintenance tasks or facility operation need to be put together and taught to the incoming facilities crew.
- *Hardware and facility components*: What is written in the specification will be compared to what is *actually in* the facility. In theory this should match perfectly. In reality it won't, and this will be the source of many discussions between the companies involved.
- *Wiring*: The building will have electrical wiring and this will come with a wiring or circuit diagram. The diagram should be checked against reality, usually via a random check of the outlets provided.
- *Inputs and outputs*: Electricity and water comes in, heat and steam goes out – and where exactly this happens will be listed in an I/O diagram. This is quite important and so should be checked thoroughly.

Once these and many other checks are complete, a long form will be filled out detailing just what was expected and what was delivered. Problems will be listed in a deviation or deficiency report, alongside the measures and deadlines associated with fixing those problems. All sides need to be happy that the facility is complete before it will be officially handed over to the pharmaceutical company.

Operational qualification (OQ)

Does it work? More accurately – can you reliably state that your facility and the equipment within it will consistently work to the required level of quality? Operational qualification can be thought of as the final check before you go live for commercial production, and much like other qualification tests it needs to be performed based on a pre-approved protocol.

Equipment and facility OQ will be similar in their end goals but will have different testing requirements. For facilities in particular, there are several important things which need to be checked and documented to make up the final qualification step:

- *Testing equipment*: You need testing equipment to test things, obviously. What might be less obvious is that the equipment used needs to be documented – what was used and when. This is to show that it was correctly calibrated according to the relevant regulations and to help trace any problems which occur later.
- *Sensors*: If there are sensors in place within the facility, then they need to be calibrated. The methods used for this calibration and, naturally, the results, will be documented as well. Similarly the effectiveness of any alarms which are attached to those sensors needs to be verified.
- *Flow rates*: Water, air, exhaust, all of these have specified rates at which they will be provided or taken away. The flow rate and the associated sensors/controls which keep this flow under control will need to be checked for accuracy and consistency.
- *Equipment*: There is process equipment (which we talk about in the next section) and facility equipment – the machinery which keeps everything in the building running. This equipment also needs to be qualified to determine that it can perform the required functions.

The facility qualification report will be based on these and other tests, and will essentially state that the building is ready for use (assuming nothing went wrong, in which case there will be a list of deficiencies and required corrective actions). Once signed by everyone involved (and there will be a lot of signatures) then the facility will be ready for process qualification.

Qualification of equipment

There is no point having a fantastically designed manufacturing process if the equipment you use is garbage. To help avoid these cases of leaking pipes, low quality water or bad autoclaving, equipment must be **qualified**. Qualification of equipment is the act of formally demonstrating that it is fit for its intended use, and it follows a very similar path to that of facility qualification.

Design qualification

The first step in qualification is to actually choose the right equipment. This is a matter of looking at the final use and then scientifically justifying the just how your piece of equipment should go about letting you achieve that. For example, if the final goal is to have a sterile, dry glass bottle for you to hold your tablets, then you should prioritise dry-sterilisation methods such as heat tunnels over 'wet' ones such as autoclaves. These requirements will be written up into what is formally known as a **User Requirements Specification**, a clear description of just what it is that you need (note that this is what you need, not what you think will solve that need. It's your requirements, not your shopping list). URS's are sent out to potential suppliers to see who can provide the best quote for supply.

Once the supplier offers are in hand, then it is time to move to design qualification. This is documented verification of whether the proposed solution can match your requirements. You will go through the requirements listed in your URS and, one at a time, determine which components of the supplier solution can actually meet that requirement. Naturally a good specification is a vital part of getting the right equipment in later, and this is why companies will spend an inordinate amount of time getting the requirements down at the start.

Design qualification usually ends with the choice of an equipment supplier, who is then contracted to provide the piece of machinery which you are looking for.

Installation qualification

The next step is to verify that your equipment was actually installed properly, that it meets the required specifications, and that everything has been properly documented. Was the piping connected correctly, did the pH meter get calibrated, is your tubing actually made with non-leaching plastic, can the sterilising tunnel hold as many bottles as you expect, etc. etc. All of these are basic things which need to be checked out before you start doing any more complicated testing.

Operational and performance qualification

Once you've verified that you have the right equipment and that it has been correctly installed, then it's time to start doing qualification tests. Operational qualification can be thought of as asking "does it work like the supplier says?" while performance qualification is more along the lines of "does it work like our process requires?"

This normally involves running the equipment under normal and 'worst case' conditions to see how it performs. We can take an autoclave as a typical piece of equipment which needs to be thoroughly qualified before use. The potential 'range' of loads could go from completely empty (sterilising the autoclave itself) through to a full load of something complex and difficult to sterilise (such as lots of long, narrow plastic tubing). A typical qualification test will look at both extremes as well as several intermediate situations. If the autoclave works according to requirements in all cases (i.e. reaches sufficient temperature and pressure, actually sterilises things as tested via bio-indicators) then it is considered to be **qualified** – ready for use.

As seen in the autoclave example, qualification tests need to match the actual conditions being used in general production. This means that the potential problems which you would expect to occur should also be simulated – it should be considered a 'worst case' check. This means that qualification will test things such as line stoppages to clear broken glassware, extended start-up procedures, holds in the process to cover shift changes, etc. etc.

As with all things in the world of pharmaceuticals, equipment qualification begins with a protocol – a document which describes what studies will be done, what the conditions will be, what the 'success' criteria will be, and at what point the equipment will be considered qualified. As the equipment in question can range from a simple set of scales up to an automated manufacturing line, the size of the protocol can similarly reach hundreds of pages in length.

Once qualification testing has been completed, all of the results will be written up into a final report. This will usually include a section explaining anything which may have gone wrong and the subsequent effect (or lack thereof) on the qualification status. These are known as deviations, and we cover them in further detail in a later section. It will also include an overall summary section stating the conclusion of the report – this will be *the* part which is read by management and so will usually be the focus of intense argument and discussion.

Once the report is written up, the qualification criteria achieved, the deviations explained and quality assurance satisfied – then the equipment can be considered qualified and ready to use.

Qualifying analytical methods

Analytical methods are like any other process-associated method, they need to be qualified or validated prior to being used. This isn't required during the initial development stages, but the expectation from regulatory authorities is that by the time you reach your commercial stage, the assays should be qualified and validated.

To clear up the terminology in use here – an assay is **qualified** when the study confirms that it is *capable* of working at the required accuracy and reliability. The assay is **validated** when it is tested against specific acceptance criteria (set in advance) to show that it *is* working at the required level. Both sets of studies involve performing a number of laboratory tests using known materials which cover the entire potential range of the assay (i.e. if you are testing for impurity levels and expect it to be between 5-10%, your assay should work between 0-15% at least). The capabilities of the assay will affect the acceptance criteria ranges in the commercial process – and in some cases you will have to find another assay if the first one is not capable of providing the precision you need.

All of these tests are documented in the respective report and the report is archived along with all of the other GMP-relevant documents. Assay qualification/validation reports are an important part of the regulatory approval process and often come up during audits. They form the evidence you need to prove that yes, you are competent in doing the testing which will be required.

Process Performance Qualification

Once you are satisfied that the equipment and facilities are up to scratch and fully qualified, it's time to move on to the process. Known as **process performance qualification**, this is a combined check of the commercial manufacturing setup using the available equipment and personnel. It's basically a test run or rehearsal for the real thing, albeit with lots and lots of extra measurements.

Process performance qualification, or PPQ, is done at commercial scale. This means that you will be working with large scale reactors, industrial sized equipment, and very impressive costs if something manages to go wrong. Because of this potential cost, pharma companies will be very careful in how they plan and execute a PPQ study – failing here means that you need to redo the entire thing, costing even more money. Once the PPQ has been successfully passed, the data and conclusions will make a large part of the submission to the regulatory authorities for final approval.

Because PPQ is a 'final exam' of sorts, it needs to match the final commercial process and commercial sizes as closely as possible (regulatory authorities will allow some variations, but you'd better have a really good reason for it).

A PPQ needs to examine three **consecutive** batches – so three batches produced one after another. The consecutive requirement is there to stop companies from producing twenty different batches, then cherry-picking the three nicest ones to report. Interestingly enough, however, *three consecutive* is usually interpreted as *three consecutive batches using the same method*: you can produce another drug or even the same drug using a different method in between those batches, as long as the batches made using the method being validated are consecutive. This is most often seen when the manufacturing line makes multiple products or when an already-approved product is being improved.

Several different approaches using manufacturing methods for Drug A – Method X is undergoing process validation, method Y has already been developed.

Batches being produced	Batch 1	Batch #2	Batch #3	Batch #4	Batch #5	Consecutive?
New product, new method:	A, X (PV)	A, X (PV)	A, X (PV)			Yes
New product, new method, cherry-picking:	A, X (PV)	A, X	A, X (PV)	A, X (PV)	A, X	No
Old product, new method:	A, Y	A,X (PV)	A, X (PV)	A, X (PV)		Yes
Old product, new method:	A, X (PV)	A, Y	A, X (PV)	A, X (PV)	A, Y	Yes
Multiple products:	A, X (PV)	Drug B	A, X (PV)	A, X (PV)	Drug B	Yes

This doesn't mean that all possible commercial scales need to be tested, you can use a **bracketing** approach if justifiable. Bracketing involves testing multiple batch sizes which cover the entire commercial range – for example, if your process is able to make between 5 and 10 kg of vaccine at a time, you can validate a 5 kg, 7 kg and a 10 kg batch. This gives you three consecutive batches as well as showing that you are able to produce at all possible scales. If you have a very simple process and amazingly persuasive data it may be possible to validate only the smaller end of the range – but this requires a lot of luck and will usually be shot down by regulatory authorities.

Because a PPQ is acting as a 'model' manufacturing run, it is much more closely observed than a normal run would be. Extra samples will be taken, more tests will be performed, and *much* more care will be used when doing all of the steps. How much extra observation is required during PPQ is a variable and depends on several factors. A simple and well-understood process will require less 'extra care' than a complex method being tried out for the first time. The additional data gathered at this stage will then (hopefully) support the use of a more limited set of measurements and controls in normal production.

Process Parameter Qualification Protocols

All studies in the pharma world start with a written protocol, a document which describes in detail just what you are going to test and how you will go about it. The PPQ Protocol forces your company to nail down its procedure and prevents the dodgier ones from pretending everything is ok when things don't work as they should. A typical well-written PPQ Protocol will include:

The manufacturing conditions
In particular things such as operating parameters, the limits which will be placed on those parameters, the raw materials and the intermediates which will be involved in the process, and (naturally) how the final product should look.

Details of data
What data are you intending to gather? How will you gather it? What analysis techniques will be used? What will the evaluation and acceptance criteria be? Putting all of this down in the protocol stops dodgy companies from choosing the best statistical technique for the data.

Tests to be performed
The tests or controls which will be performed at each process step and the acceptance criteria which will be associated with each of those tests. For example, if you are going to measure the pH of your buffer after mixing, you'd better have a range that those values should be in. It was previously possible to get away with simply 'reporting the value' of a control, but these days this is considered to be Bad Form and is rarely acceptable. Tests will be set for process steps during manufacture (in-process controls) or for the finished product (known as release testing).

The sampling plan
The sampling plan shows in detail where in the process samples will be taken, how they will be taken, how many will be taken, and how often they will be taken. The number of samples which need to be taken is usually determined based on the risk profile of the attribute in question as well as statistical considerations – a highly important and highly variable factor will need a lot of samples to ensure that it is measured accurately and with statistical reliability, a boring and reliable one will not. As you would expect, the number of samples being taken during PPQ runs is much higher than the number taken during routine production.

Deviations

Eventually (inevitably) something will go wrong. It may be an in-process control going outside the acceptance limits, it may be a value trending in the wrong direction, it may be the power going out and halting production for a day or two. Just as important as good manufacturing is being able to deal with problems as they come up. The protocol doesn't need to explicitly state this (problems will tend to be unique, after all), but it does need to indicate that the quality group has a system in place for these problems if necessary. You also need to clearly state how you will assess the impact of the deviation on the overall success of the process qualification – no ignoring bad data or mistakes if you don't like what happened!

Facilities and Equipment

Generally all of the facilities and equipment being used in the PPQ have been qualified by this point. If so, then this part of the protocol will reference those studies (all companies should have a reliable document numbering system to help in cross-referencing). If it hasn't been done already, then now is the time to organise the studies – these are a vital part of the overall validation process.

Analytical methods

Just as the equipment being used needs to be qualified beforehand, so to do the analytical methods being used to actually take your measurements. We've covered this in a previous section, but remember that your protocol needs to list the methods which will be used, what SOPs are used to run that method, and which qualification studies have been performed to show that it works.

Master batch records (MBRs)

Production during the PPQ runs will be performed according to **master batch records**, or MBRs. The MBR is a general instruction document which acts as a giant checklist for the entire manufacturing process. It specifies what is to be done and which tests need to be performed, then contains check boxes to indicate that the steps have been performed and for test results to be included. All of this information is entered (usually by hand) as the production run continues, making a contemporaneous record of everything that happened. Once production is complete QA will check and sign the filled out document, which is known as an **executed**

batch record. This is a GMP document and so must be archived correctly, copies will also be needed by various people for the PPQ report and later regulatory filings.

The process performance qualification MBR can be hundreds of pages long, and so is often included as an attachment to the protocol document rather than within the protocol itself. Nonetheless, it must be reviewed and approved alongside the protocol prior to production going ahead. Reviewed in detail too – you would be amazed at how many very obvious mistakes make their way through several reviewers to be caught just before production begins.

Review and approval
Document review! It can be amazingly boring but is nonetheless extremely important, in particular because of the aforementioned mistakes. Many parts of the pharma industry work using signed documents, often PDFs – these are useful for ensuring that nothing changes but a pain to use as a source for copying information out of. As a result there will often be situations where minor errors will creep in to slightly modify the new document. Often this is misspelled words and the like but it is quite possible that your CIPC acceptance criteria will be copied in wrong. Document review helps to fix this.

Why is this so important? Once signed, the protocol becomes 'fixed' – it can no longer be changed without a lot of effort and problems. This means that any mistakes you miss during the review phase will be very, very difficult to fix. While this is not particularly important for spelling errors or typos, making a mistake in the planned acceptance criteria or the date of the last equipment qualification will cause you huge problems in an audit or other regulatory inspection. PPQ protocols are usually reviewed by the production, quality, and regulatory departments, although experts from other areas will also often be involved. Once everyone is happy, the document will be finalised, signed, and put into the archive – it is now Very Official Indeed and must be followed.

If you do end up in a document review cycle, there are a couple of important things you should watch out for:

- First and most importantly - the deadline. "Well obviously", you're probably thinking, "that will be in the email I get". Sadly it's not quite that easy, as many people will either not notice the deadline (usually because it wasn't clearly highlighted in the email) or simply lose track of it amongst all their other work. If you are the reviewer, look carefully for the deadline and determine how it fits in with your other work. If you are the document author, *call the reviewers first*. The review process is much smoother when you speak to people about their workload and expectations *before* they get the 'please review' email.

- Next, what are you reviewing? You should be reviewing the 'almost final draft' – don't waste your time on half-written documents. You also need to know who else is doing a review, and what order it will be in. Documents with multiple reviewers will often change immensely in a short period of time, so you will need to make a couple of checks (this, incidentally, is why it is usually better to have one or two reviewers rather than a group of them).

- Now onto the review itself. Does the document and the conclusions make sense? Based on your expertise, do you see any problems which may occur or areas where more care should be given? Are all the acronyms and jargon explained correctly, preferably in a table or glossary at the back? Are the supporting studies referenced clearly, with working hyperlinks and the like? There should be a **source document** for everything you are looking at. A process qualification protocol does not just come out of nowhere, the company spent a lot of money getting everything worked out beforehand. The document in your hands should match what is being said in those source documents – or at the very least have a reason for being different.

- *Use Track Changes mode!* There is a special place in hell for people who make extensive changes to documents and then tell you that they've "fixed a few things".

- Found things which need to be fixed or changed? Minor changes or typos should be fixed directly in the document (using track changes, obviously). If there are larger changes you want made, then you should leave comments and discuss them with the author. You should always know *why* you want these changes. There will often be long discussions about the inclusion or exclusion of certain information, tests, limits, and many other things – with different experts having wildly different wishes about the document content. Be clear as to what you want and why, because this will help you win out in these arguments.

- Once all the arguments are over, a final version will be sent out for signing. Take the final version and compare it against your previously reviewed version. Have they fixed up the mistakes you pointed out and made the corrections that you wanted? If not, why not? Are there still issues open? Remember that by signing a document you take partial responsibility for it, which in turn means that you will be blamed when something goes wrong. Don't sign unless you are happy with it!

Performing the process parameter qualification

The protocol has been written, reviewed, argued over, and finally approved – this means it is finally time to actually do *manufacture* something. Once again, it's considered to be a trial run for the real thing – you will follow the commercial manufacturing process and use the typical, routine procedures, performed by the typical, routine employees. All the conditions should be as normal as possible, even the parameters of the surrounding facility.

You want to be doing the study *exactly* as it was written in the protocol. Theoretically, this should be an easy part of the process. You've got your experience at large scale from producing Tech Batches, you have all of this knowledge from development work, you have a well-written and thoroughly-planned-out protocol and master batch record. It should now just be a matter of making the PPQ batches, just like you said you would do.

Sadly, it is never quite that simple. Every PPQ run is different in just how it becomes 'complicated', but there are a couple of different reasons which you will see fairly often:

- Things will go wrong
- Changes will need to be made
- The data interpretation will be controversial
- Writing the report will be controversial

Let's have a look at them in a bit more detail.

Things will go wrong

Despite all of your care and good intentions, something will always manage to go wrong. It may be something minor (a vial was dropped and so we had to pause the process for an hour to clean up), it may be something major (we discovered that the analytical group can't actually measure the IPC) – in the end something will go wrong.

This leads to two very important terms in the world of pharmaceutical manufacturing, namely **deviations** and **deviations to the protocol**. *Deviations* occur when something goes wrong with the process – an out of specification event, a strange observation during the study, monkeys climbing in through the window. *Deviations to the protocol* occur when the process followed during validation doesn't match up with the one laid down in the PPQ protocol – modifications to IPC testing methods, changes in stirring time, the incorrect number of monkeys climbing in, etc.

These deviations will be investigated by the quality assurance group and the **root causes** (i.e. underlying causes) found. The findings from these investigations will in turn lead to **CAPAs** which try to prevent the error happening again. All of these deviations, root causes and CAPAs will need to be included in the PPQ report – indeed the report cannot be signed until *all* of the associated deviations have been closed.

Changes will need to be made

Despite all of the care taken during the review process, there will often be some sort of change which needs to be done. This can be due to changed requirements, the desire to add more testing points, or as the response to some problem during manufacture. How difficult this is to deal with depends on when the change needs to be made.

A change *prior to starting* the PPQ is not too problematic, as it simply requires that a new version of the protocol be put together and signed. The new version will include a change history section which details the change which was made (and the reason for doing so) – this allows later reviewers to follow what was done. A change *during* the PPQ is much trickier. It is guaranteed to trigger a deviation of some sort and should only be done in cases where there is both a solid reason for doing the change and an excellent scientific justification for the change being made. As you can imagine, neither of these are particularly good options – so review the protocol well beforehand!

The data interpretation will be controversial

Numbers are fairly reliable – if a batch IPC gives you a pH reading of 3.2, then it's a fair bet that you are dealing with a pH of 3.2. The problem comes in the interpretation of these numbers, in particular when deciding how they relate to the acceptance criteria and other applicable limits. Each drug will have *registered* limits for IPCs and release tests, these are included in the dossier and registered with regulatory authorities and thus the measured value *must* be within these limits. They will also have *internal* limits, much tighter limits which are not registered but which force the company to manufacture to a higher standard – this in turn ensures that they will always stay within the registered limits.

The question then is which limits should be applied when rating the success of the PPQ batches. The internal limits, which the company should conform to? Or the registered limits, which they *must* conform to? Life gets even more complicated with products registered in multiple countries, as the limits required by the United States' FDA may be very different to those required by Cuba's CECMED. To avoid this most companies will use their own internal limits, but this can lead to a lot of discussion when deciding if a measurement is really out of specification or not.

Writing the report will be controversial

The process performance qualification report should summarise the data which was collected, following the outline which was set up by the protocol itself. As mentioned before, if anything came up unexpectedly (i.e. deviations) and required additional or alternate testing (i.e. deviations to the protocol), this should also be described.

The last part of the PPQ report, arguably the most important part, is the conclusion. This section states *whether the process validation was successful or not*. How do you know if it was successful? The authors and reviewers need to look at the protocol and answer a couple of tricky questions, namely: Did the process fulfil *all* the conditions which were set out in the protocol? Is the process considered 'under control'? These questions are tricky because answering them requires looking at all of the data which has been generated over the course of the qualification. The measurements need to be in-spec, reliable, and cover the important parts of your manufacturing process, the inevitable deviations which occurred need to be explained and determined to be irrelevant to the validation status. If everyone decides that this is the case, then the validation is considered successful and everyone will breathe a sigh of relief before going out for a party.

On the other hand, the data may not support your intentions – the validation may have been failed. Now things get complicated, as you will need even more data showing just *what* exactly went wrong, *why* it went wrong, and *how* you are going to change the process so that it will not go wrong again. Deviation investigations and corresponding arguments for a failed PPQ study will cause immense amounts of stress for all concerned. Not only is a failed PPQ a major red flag during the approval process, the entire process will need to be repeated – as each study uses commercial-scale batches this can get very expensive very quickly.

Assuming everything worked out well, then the document will be finished and it is time for all the parties involved to review and sign the document. Each group (regulatory, quality, etc.) will have their own fields of expertise and their own idea of what the 'ideal' report is. This means that you will encounter many long discussions over numerous minor factors in the document, particularly points which you personally will think are pointless. Ironically, these will often turn out to be *really* important in about 90% of cases (the other 10% will be a waste of time for all involved). As with all document review, multiple viewpoints are required to end up with a perfect final document – so don't complain too much. Eventually everyone will be satisfied (or at least equally annoyed) and the document will be signed.

Congratulations! You have successfully validated your process.

Continuing Process Validation

Your successful PPQ was the last bit of evidence that the regulatory affairs department needed to get the drug approved. Hooray! Now you just need to produce batches for the market and use the profits to buy a small island in the Caribbean. Right? Wrong! Just because you can make three batches in a row correctly doesn't guarantee that you can continue to do so in the future. Thus you now get to enter the world of **continuing process validation (CPV)**, where the quest for better quality never stops.

Continuing process validation, or CPV, aims to show that the process which you validated *remains* validated. Just as important as having a successful manufacturing run is having each following run be equally successful. To ensure this, you need to follow how their parameters **trend**, or change over time. Is the purity of your intermediate product decreasing slowly, even though it's still inside the specifications? You should know, because it's a sign of problems further down the line. Regulatory authorities like to see plans for long-term trending of measurements, which plays a very important role in Continuing Process Validation.

Trending data requires that you first gather enough data – data which is taken from all of the commercial batches being produced. The official requirement is that you are able to know enough about the process that you can spot unwanted variations, determine where they are coming from, anticipate when it might happen again, and then adjust your methods to keep everything under control. In practice this means establishing a data collection and analysis program, one which covers in-process controls, release data, process observations and even raw materials.

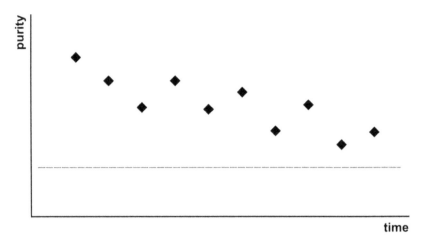

Trending is important for spotting problems which have not yet occurred. A series of batches are produced and the final purity measured (black diamonds). Although all of the results are above the required specification (dotted line) a clear trend towards an out-of-specification event is visible. Trending helps you avoid this problem before it occurs.

Normally a statistician will be brought on-board to help determine when and how often data points should be collected (each extra test performed costs money, so the company wants to balance quality control and budget needs). They will also be involved in analysing the data, tracking long term trends or spotting potential out-of-specification problems. The overall aim is to clearly demonstrate that your usual routine keeps everything under control and consistently makes a high-quality drug.

As mentioned, this usually involves statistics. Although non-statistical based methods do exist, regulatory authorities tend to dislike them – they often lead to over-reaction to isolated incidents or under-reaction to consistent, low-level problems.

CPV starts directly after commercial manufacturing begins, at which point you will typically only have made a couple of commercial-scale batches (technical and process validation batches, usually). This means that, realistically, you have no idea what weird little things will possibly pop up in your process and cause variation in the final product. *Because* there is such a lack of knowledge at the start, regulatory authorities will expect you to monitor your normal production with almost the same level of care as you did during PPQ. As you get more and more batches made you will begin to spot which process steps vary more than the others – these are the steps which need a closer look. Thus as the company gets more experience with the process, they are expected to increase and decrease their control of the process steps to match their increasing knowledge.

The plus side of gathering all of this data is that you can spot points in the process which could be improved. As commercial-scale manufacturing involves millions of dollars each time a batch is produced, even a small boost in efficiency can make a huge difference in profitability. Actually getting these changes implemented is slightly trickier, however, as it then falls into the realm of Change Control.

Change Control

Change Control is a general term for the process of making a change to a system – in particular the act of making a change in a controlled and reasoned way (i.e. not just changing everything because it seemed like a good idea). Change control generally has a number of defined steps:

- *Origination*: Someone in the organisation will request that a change is necessary for a process, a piece of equipment, a building, even within the organisation itself. This is recorded within a dedicated system as a new **change request**.
- *Evaluation*: A number of experts will assess the impact that this proposed change would have on their specific areas of expertise. They will determine what sort of activities would be involved, how much time and money they would cost, and who would be performing them. This information is taken by a management council who will then make the decision on whether the change should be implemented or not. As you'd expect, many changes are cancelled at this stage due to impracticality.
- *Planning and prerequisites*: The change has been approved, but there are a number of things which need to be done first. This stage will involve planning of the change, determining the prerequisite actions and who will do them, and working out

overall timelines. Expect a number of meetings where everyone argues over how the entire thing will be done at this point.

- *Implementation*: Time to actually make the change! The planned out actions will be performed and the newly changed system will be unveiled for the world to see.
- *Close-out*: Hopefully the change has actually been worthwhile and has achieved the desired aim. Once everything required has been done, (this can take many years) then it is time to close the change.

GMP guidelines require that all companies have a reliable change control system, one which is able to document change requests, evaluations, and implementations. Although some workplaces will still use a paper-based system, your employer will almost certainly do this via an electronic database/change-management system at your company. The details will vary, but the basic goal of the IT implementation is to force everyone to follow the correct process and then record everything which was done. The system in place is of particular interest to auditors and they will often go through random changes or quality events trying to find a mistake or out-of-place entry.

Change control will affect many different fields within a pharmaceutical company, but you will find that certain groups will spend more of their time dealing with it. Many of the change requests will originate in production, raw materials, or quality departments (as these are all the most likely to either think of a process improvement or notice something which needs changing). Quality assurance will almost always be involved, they will usually have the final say on whether GMP-relevant changes should go through. Regulatory affairs will find themselves either evaluating the changes or trying to get approval for the ones which have been approved (a process which can take years if you are unlucky). Various other departments will also find themselves sucked in depending on the scope of the change request. In general, however, if you end up working in a production-focused area then you will almost certainly end up dealing with change control.

Appendix: Risk Assessment Methods

There are a number of tools which have been put together to help out with risk management, each of which have different focuses and different advantages. This section will cover some of the most popular options which you'll run into.

Failure Mode Effects Analysis (FMEA)

If there is one risk assessment which you are going to run into, it's going to be FMEA. As the name suggests, FMEA focuses on what are known as **failure modes** – states in which something specific has gone wrong. The key term here is *specific* – there is no failure mode such as 'the pH of the buffer is wrong', instead it is broken down into clearly distinct failure modes: 'the pH is too high' and 'the pH is too low'. Even though these affect the same system, they may be due to different causes and so need to have separate assessments.

FMEA is a systematic approach, using the idea that you can break any process down into a number of isolated steps and systems. Each of these components are then examined carefully and all of the potential failure modes are listed in a formal table normally known as a 'FMEA worksheet'. The consequences of each failure mode are also listed in the sheet, as is any information regarding the root cause of such a problem.

At this point the assessment part of the process comes in – you will determine the probability, severity, and detectability of each different failure mode. These are given different ratings based on their assessment as seen in the table below. How you actually assign these values can be qualitative ("it happens about this often") or quantitative ("the chance of occurrence is 13%"), depending on how much information is available to you at the time.

This table shows the typical FMEA assessment factors

Probability	Severity	Detectability
1. Extremely unlikely	1. No real effect	1. Certain to be noticed
2. Remote	2. Very minor (no damage or injuries)	2. Almost certain
3. Occasional	3. Minor (some damage or injuries)	3. High chance to be noticed
4. Reasonably likely	4. Critical (severe damage, maybe someone dies)	4. Moderate chance
5. Inevitable	5. Catastrophic (a lot of people die)	5. Low chance
		6. You'll never notice

Once these ratings have been assigned you can then combine them to determine criticality. This sounds complicated but is basically a matter of plotting your ratings out on a **risk matrix** grid like the one below. As you can see, the acceptability of any particular failure mode is directly linked to *how bad it is* and *how often it will happen*. There are many types of risk matrix grids in use and the particular one you use is dependent on the specifics of the project and the failure modes being looked at.

An example risk matrix for failure modes with a Moderate Chance of being detected

		Severity			
		1	**2**	**3**	**4**	**5**
Probability	**1**	Irrelevant	Low	Low	Moderate	High
	2	Low	Low	Moderate	High	Unacceptable
	3	Low	Low	Moderate	High	Unacceptable
	4	Low	Moderate	High	Unacceptable	Unacceptable
	5	Moderate	High	Unacceptable	Unacceptable	Obviously bad

Based on this risk rating you then decide which failure modes are acceptable and which aren't, thus setting the stage for prioritising risk reduction efforts. The risk matrix step blurs the line between FMEA and the ploddingly-named Failure Modes Effects and Criticality Analysis

(FMECA). FMECA can be thought of as being *like* FMEA, but adding information on how critical the failure is to the mix (it's more complicated than this, but we won't get bogged down in details).

In any case, all of this information will go towards filling in a ridiculously large table covering failure modes and risks for your entire process. This table in turn makes up the majority of your risk assessment report, with an introduction and conclusions section to help bring the discoveries together.

As with all processes, FMEA has pluses and minuses. It's an excellent method for breaking down complex systems and determining the important points of failure (and the associated consequences). However the scope of the analysis needs to be carefully chosen to ensure that you're rating problems with the right process and not accidentally skipping something in a related-but-not-in-scope system. Similarly, the ease of rating failure modes means that it is simple to take up and use, but is heavily dependent on the quality of the risk matrix – is a medium severity just as bad as a medium probability? Probably not, so your risk matrix has to take this into account. This, coupled to the horrifically ugly complexity of a large FMEA worksheet, tends to discourage immediate uptake.

Nonetheless FMEA remains the oldest and most popular of the structure risk assessment techniques. If you are going to be involved in risk assessment, you are probably going to be working with FMEA, so you should get used to the approach.

Fault Tree Analysis

Fault tree analysis (FTA) is a top-down method for determining risks, basically breaking down an event into combinations of lower-level events linked by Boolean statements to figure out the risks at the top level. Actually performing an FTA is done according to a series of steps, as detailed below:

Step One: Define the unwanted outcome

What could go wrong? Much like the failure mode from before, we try to list each possible problem which could occur during the process (this is known as an *undesired event*). Each of these events acts as the starting point for a Fault Tree Analysis – one event per tree.

Step Two: Understand what is going on

What are the different causes which could cause the unwanted problem? How likely are they to occur? How do they interact with each other, so changes at one point will modify other possibilities?

Step Three: Make a fault tree

We assume that at this point we know enough about our event to link up the various causes into a tree – the idea being that higher-level causes will be closer to the starting point, branching down via OR and AND combinations to root level causes.

Step Four: Analyse that tree

Are there any parts of the tree which have an oversized impact on the overall risk? Where can your risk-management approach be improved?

Step Five: Improve your process

Go out and implement those improvements which you spotted in step four.

Let's explain this in better detail by looking at a failure state using our example drug, the cunningly named Examplifen. We've recently discovered that the final drug product may have trace contamination from aspirin (acetylsalicylic acid), which is produced on the same manufacturing line. Contamination is, naturally, a bad thing for drugs, and so we would like to prevent this. We begin a fault tree using the contaminated state as the beginning node (as we can see below).

The first layer of the tree covers the most general contributing causes. One of these is very basic – the aspirin and Examplifen must be produced with the same equipment and on the same line for cross-contamination to occur. The second contributing cause is that the equipment cleaning wasn't performed correctly. Note that both of these need to occur before we get cross-contamination: bad cleaning won't allow aspirin in if there is no aspirin present in the system; if aspirin is produced on the same line then effective cleaning would prevent cross-contamination. As both of these events are necessary for our failure state, they are marked with an AND.

We move to the next layer, with slightly more specific contributing factors. One option is that the cleaning wasn't performed adequately, the other is that the cleaning simply wasn't performed at all. These are independent and exclusive of another, so we can mark them with an OR. 'Cleaning wasn't performed at all' could be broken down into several further sub-factors (e.g. the cleaner was sick, someone forgot the protocol, an ice-cream truck drove by and everyone left), but for the moment we will consider this the **possible fault** and it terminates that branch of the tree.

On the other side, Cleaning Wasn't Performed Adequately can be broken down into a number of sub-factors which can independently lead to the layer above – hence we have another OR choice. These three sub-factors all represent different ways in which the cleaning may not have been performed correctly, and so make up the next set of possible faults.

In this way we now have a number of possible contributing factors which can lead to our original failure state, we can use this tree to identify the best place to do our risk management. In this case, for example, we can spend time and money improving the equipment cleaning processes – this requires us to work on *all* of the contributing factors (as any of them may lead to the final fault). Or we can just stop using shared equipment between aspirin and Examplifen – this is an absolute requirement for our fault to occur and so we can attack the problem at the highest possible level.

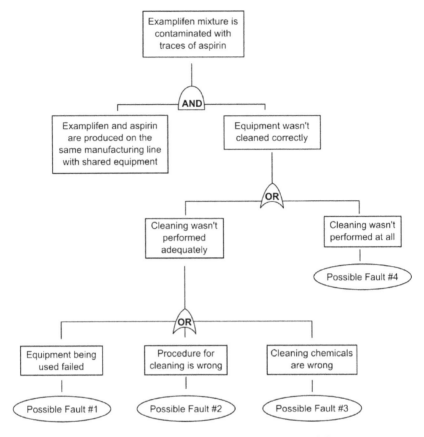

An example fault tree analysis performed for a single failure state – where our drug mixture is contaminated with traces of another drug.

FTA is a top-down system, which means that you focus on a single potential problem and then determine the causes for that fault. Because of this FTA is very useful for looking at how faults combine to affect a single system, but is bad at identifying all the faults which may occur across an entire process – this is better done using a bottom-up process such as FMEA. In practice a really thorough risk assessment will use both FMEA and FTA approaches to combine breadth and depth of assessment.

Hazard Analysis and Critical Control Points

HACCP is predominantly used in the food and beverage industry as a systematic way to identify and prevent potential contamination points, having been developed in the 1960s as a way to ensure that the food sent into space alongside astronauts was as safe as could be. From there the uptake of HACCP has spread to cover many foodstuffs (it is a requirement for several 'high-risk' foodstuffs in the US such as meat and juice) and from there into other industries. HACCP is an approach which is most useful when the process is very well understood, as this allows the 'critical points' to be identified and dealt with.

The basic process is as follows:

1) *Conduct a hazard analysis*: What hazards could affect the quality of the final drug product? What preventative measures could be set into place to prevent these hazards?
2) *Determine critical control points*: These are the points at which an IPC can be inserted to prevent or reduce the hazards you've identified.
3) *Establish critical limits*: These are the limits of variation which you will accept in your IPC before you start having quality issues.
4) *Establish monitoring processes*: How will you actually monitor that your IPC is within the right limits?
5) *Establish corrective actions*: What happens when the variation goes outside the limits (i.e. a deviation)? These should be written down clearly.
6) *Establish a validation system*: You then have to verify or validate that your system is working as it should be.
7) *Establish a record-keeping system*: All manufacturing runs and associated values need to be recorded and stored.

As you can see, despite the origins in food safety this is essentially the same as the development of a manufacturing process for drug products. The specific formal requirements do differ slightly, but in general a pharma company which is developing according to GDevP/GMP will not have a problem performing a HACCP analysis.

Appendix: Glossary of Terms

Term	Meaning
Absolute risk	The chance that something will happen – a manufacturing process where something goes wrong in one of every ten runs has an absolute risk of 10% (you can also state this as 1 in 10 or a probability of 0.1). Not to be confused with *relative risk*.
Acceptance criteria	Criteria or specifications which must be met. These can apply to finished products, intermediate measurements, or raw materials.
Active Air Monitoring	Using what is essentially a fancy vacuum cleaner to actively pull air across an agar plate, so as to determine how many microbes are floating around in your critical zone
Airlock	Just like on a spaceship, this is a room with two doors designed to keep one side (the outside environment) separate from the other (the clean zone). There is usually a pressure-difference across the airlock to make sure microbes are blown away from the clean zone.
Alert level	The point at which your monitoring system says 'there *may* be a problem', and which leads to closer examination of the process.
Action Level	The point (higher than an alert level) at which there *is* a problem, and where you *have to* perform an investigation.
Active Pharmaceutical Ingredient (API)	The part of your final product which has an actual pharmaceutical effect.
Aseptic Production	Producing your drug in an extremely sterile environment, as opposed to terminal sterilisation. This is the usual approach taken by biological drugs.
Aseptic Processing Facility	The part of the building where aseptic production is performed, separated from the outside environment by physical barriers, air filters, etc.

Bioburden	The number of microbes which are found on an item before sterilisation (or after, depending on your process).
Biological Indicator	A group of test microbes which are placed inside the equipment while testing the sterilisation process – if the sterilisation has worked, these microbes should also be dead.
Blow-Fill-Seal (BFS)	A manufacturing technique where plastic containers are extruded, blown into shape, filled with sterile liquids, and then sealed – all without exposure to the outside environment.
Bracketing	Producing several batches of different sizes to cover the entirety of the range of sizes which you want to produce commercially. Common practice in process validation.
Bubble-Point Testing	Determining the presence of defects in a liquid filter by measuring the pressure needed to force bubbles through it.
Change request	The official request to make a change to a currently running system, followed by a change evaluation and acceptance process. A vital part of GMP requirements.
Corrective and Preventative Action (CAPA)	The outcome of a deviation investigation should be a set of actions that you will take to prevent the problem happening again – CAPAs.
Clean Zone	A subset of the aseptic facility, this is a region or room which has a set cleanliness level and which has been verified as achieving that level.
Clinical batch	Drug product batches which are produced for clinical studies.
Closure	Your drug is filled into a container, which is then sealed with a closure – think of lids, rubber stoppers, etc.
Colony forming unit (CFU)	After plating out a solution onto agar, a single microbe will grow into one small colony. This is a single CFU, and is usually considered to be 'one bacteria' (though it can be more in some species which 'stick' to each other).
Commercial manufacturing process	The manufacturing process, the one which you use to make your final, ready-for-sale drug product. Regulatory authorities usually make a

	distinction between the commercial manufacturing process and the one used to make material for clinical trials (which occurs earlier in the application process and as such is usually less optimised).
Component	The term used to refer to the ingredients used to manufacture the final drug product – this can refer to the API, excipients, even the water for injection.
Compounding	The combination of individual ingredients to create the final drug product mixture.
Concurrent release	Releasing a product batch (i.e. allowing it to go to market) which was made during qualification before the entire qualification study has been completed. This is possible in certain circumstances but you must be really, really sure that all the criteria have been met.
Consecutive	Consecutive batches are those which are produced using the same process one after another.
Contact Plate	Specially designed agar plates which can be directly pressed against surfaces to pick up any microbes which may be there.
Container	The sterile drug is filled into a container, such as a vial, a pre-filled syringe, etc. which is then sealed with a closure.
Contaminant	Something which should not be in the product and renders it unfit for use.
Continuing process verification	Constant checks performed during normal manufacturing to ensure that your process remains under control.
Control strategy	The overall approach which you use to keep the quality of your product under control.
Critical Surface	Any surface which may directly contact the sterile product or container and thus one which needs to be carefully sterilised.
Critical Area / Critical Zone	The zone in which the sterile drug product and containers are exposed to the air, and thus the part with the highest requirements for cleanliness.

Criticality	How important something is – usually in the sense of how likely it is to affect the quality of the final drug product.
Design of experiments	An organised approach in which a number of experiments are performed and statistical tools used to determine the most important factors.
Design space	A way of combining a number of process parameters and limits to create an 'area' in which you can work without worrying about quality problems.
Detectability	How likely you are to actually notice that a problem or hazard has occurred. A normally colourless drug suddenly turning orange is a fairly obvious and thus highly detectable hazard, a slight drift in ionic strength will be more difficult to detect.
Deviation	Basically: When something goes wrong, requiring an investigation.
Deviation to the protocol	When something goes wrong, because you didn't do what the protocol said you would do.
Disinfection	The process of killing everything which may be living on the surfaces of your clean room.
Dossier	The regulatory dossier is a summary of all the information gathered on a drug, the manufacturing process, the control strategy, and various other pieces of information. Often thousands of pages in length, it is submitted to the healthcare authority to obtain permission to market a drug in a particular country. Once approved, it forms a binding contract with the country and thus any changes to the process must be reported.
Drug	Technically we should say 'pharmaceutical' but no-one ever does (even the FDA). Fit in to the field by saying 'drug' instead.
D-value	The time (in minutes) which a sterilisation process needs to reduce the number of microorganisms by 90% (i.e. 1-log).
Efficiency Testing (filters)	Testing a filter to see how effective it is at removing microbes.

Excipients	Everything which is present in the final drug product which is not the active pharmaceutical ingredient.
Empty Load	A sterilisation run without any contents to be sterilised, especially used for validation of the autoclave.
Endotoxin	A molecule such as lipopolysaccharides which are found in bacterial cell walls, and which lead to very dangerous immune reactions when introduced to the body.
Environmental monitoring	Checking for the presence of particles, microbes, unwanted air-flow, etc.
Executed Batch Record	The 'filled in' batch record, with all of the times, measurements, observations, etc. clearly recorded and signed off. This is *the* official record of the production process.
Failure modes	The many ways in which something might go wrong.
Failure mode effects analysis	Looking at how the many ways in which things may go wrong can cause many other things to go wrong.
Good Manufacturing Practice (GMP)	A best-practice approach to ensure that drugs are manufactured to a consistently high quality standard.
Hazard	A hazard is simply a potential source of harm, so something which may cause problems with drug quality, availability, etc. Your risk assessment will focus on the different hazards and their potential effects.
Heat Penetration Study	Testing the sterilisation process when loaded with typical items, to check whether tightly-packed or sealed objects are still sterilised correctly.
HEPA Filter	'High efficiency particulate air' filter, a filter which removes 99.97% of all particles that are $0.3\mu m$ and larger.
Hold Time	A potential pause in the production process, it is not always required but provides time to fix any problems which may occur.
Impurity	Something which reduces the purity of the drug. Unlike contaminants, these are often

	degradation products of the API or remnants from the manufacturing process.
Intervention	The process of doing 'anything' which is not part of the normal production process within the aseptic zone.
In-process control	A test or measurement to keep track of a variable parameter.
Isolator	A large protective shell which protects the production line from the surrounding clean-room. It can be closed or open, depending on requirements, and has the same cleanliness as a critical zone.
Leak Testing (filters)	In-place testing of the filter and its housing to ensure that it is working correctly.
Manufacturing Campaign	Manufacturing a series of batches in a period of time.
Mapping Study	A study to check if the sterilising equipment and process leads to equal temperature and pressure throughout the entire system.
Master Batch Record	The not-yet-filled-out record containing instructions for all steps in the production process.
Media Fill Study	A 'practice' run of your normal production/filling process using microbial growth media, to check whether everything comes out sterile.
Membrane Filter	A polymer membrane with tiny pores that block microbes/particles. Can be used for liquids or high-pressure gases.
Passive Air Monitoring	Checking for the presence of airborne microbes by leaving an open agar plate out for a few hours.
Performance indicators / performance metrics	Literally something that you measure to determine how well you are performing. In this context it usually refers to quality-related attributes.
Possible fault	Fault tree analysis
Process analytical technology	An approach to controlling critical quality attributes with a focus on real-time testing and adaptation.

Process capability	This is the term used to describe the ability of a process to make a final product which fulfils the desired requirements. It is often stated in statistical terms, in particular looking at the likelihood that a product will not meet the requirements (i.e. will be out of specification).
Process characterisation	The act of performing numerous tests on your manufacturing process to see what variations lead to which changes in the final product or result.
Process control	The general term for all of the work which is needed to keep a process running correctly.
Process design	This is the act of taking everything you know about your manufacturing process, from scale up and initial development studies, and using it to design a reliable commercial-scale method.
Process parameter	A variable in the process.
Process performance qualification	A formal confirmation that your process is able to reproducibly make high-quality drugs. This is usually the 'last step' required before regulatory approval is obtained and full manufacturing can begin.
Process step	Also known as a Unit Operation, this is a single piece of the overall manufacturing process.
Process validation	The overall approach by which you gather data and evaluate it to show that your commercial process is pretty damn good. It comprises a number of steps which we have explained in this book, but it can be generally described as: proving (and documenting) that your process is highly reproducible and leads to a high-quality final product.
Product lifecycle	Like it sounds, this is the lifecycle of the product – from newborn beginnings as a potential development candidate through to the sales and ongoing testing of the finished product post-approval.
Protocol	A general term for the document which describes what you are about to do and how you are about to do it. Protocols are a vital tool for

	keeping all sorts of studies on track and are a basic GMP requirement.
Qualified scale-down model	A 'mini' version of the manufacturing process which has been *qualified* – determined via testing to act in a representative way to the main process.
Quality	Quality is a fairly nebulous yet extremely important factor in the world of pharmaceuticals. At its most basic it can be considered the degree to which the properties of a product or system match the requirements. In practice it is a number of overlapping fields all of which aim to end up with the best final product possible.
Quality attributes	Any type of property or characteristic that needs to be within a certain range – if you go outside the range, the quality of your final product suffers
Quality by design	Is the principle of setting up your controls and specifications so as to reach certain specified objectives, as opposed to testing a bunch of test batches and then basing your specifications off their variability (this is sometimes known as 'quality by QC testing').
Qualification	Testing and documenting that the equipment is capable of acting as it should.
Qualified	Formally tested and shown to be suitable for the task.
Relative risk	Relative risk is the chance that something will happen – *in comparison* to another situation. In other words, a two-fold increase in relative risk means that the second risk is twice as likely to occur as the first. Not to be confused with *absolute risk*.
Release	Somewhat like homing pigeons, drugs are released from the factory into the marketplace where they can be sold. Unlike homing pigeons, no-one wants the batch to come back.
Release criteria	The criteria or specifications which need to be fulfilled in order for the batch to be released to the next step.

Release testing	The testing which is done to check those release criteria.
Report	The flip side to a protocol. A report brings all of the data obtained during a study together and makes a final judgement on the outcome. Also a vital component of GMP.
Residual solvent	Left-behind traces of the solvents used to produce the drug ingredients.
Risk	One of those words which you will hear a lot, it represents a combination of how likely something is to cause harm with the severity of the harm that ensues.
Risk acceptance	The decision to say "yes, this particular risk is something which we can live with" (this is *definitely* the sort of decision where you really want to be sure that you've done the analysis right). Risk acceptance is the contrasting option to risk reduction.
Risk analysis	Taking a hazard and determining what the associated risk could be.
Risk assessment	This is the overall process of looking at all the potential hazards in a given process, analysing the level of risk associated with them, and then evaluating what your response will be.
Risk communication	It's pointless discovering an unknown major risk in your process if you then just wander off to a bar and forget about it. Risks need to be communicated to people who can do something about them – this usually means management of some sort.
Risk control	The process of controlling your risks.
Risk evaluation	How much of a problem is your risk? Evaluation is about using systematic scales to figure out just how bad the situation actually is.
Risk identification	How do you evaluate something if you don't know what it is? The first step in the process is to identify all of the potential risks associated with your process.
Risk management	This is a general term for all of the actions which you take to identify, assess, and control the various risks associated with your work.

Risk matrix	A grid which helps you to determine just how risky your identified hazard is, based on the probability/severity ratings. Used extensively in approaches such as FMEA.
Risk reduction	The act of doing something which will reduce the riskiness of your process: think of adding new tests or changing methods. Usually you can't completely get rid of a risk, you can only reduce it to an acceptable level – this is where risk acceptance comes in.
Root cause	The underlying reason why something went wrong. Usually quite difficult to determine and so deviation investigators usually need to go down through multiple levels of cause-and-effect.
Scale up	Moving from one manufacturing scale to a larger one.
Scale down model	A 'mini' version of the manufacturing process.
Source document	Many reports will be based on other reports, or will include data drawn from other reports. The source document is the original source of this data.
Specification	The requirements or criteria which something must fulfil.
State of Control	When a process is in a 'state of control' it can be reliably assumed that the end product will have a necessary level of quality. An adequate state of control comes from the various tests and in-process checks that occur before and during manufacturing.
Sterilisation	Killing everything! Everything microbial, at least.
Sterilisation Assurance Level	The probability that a sterilisation run will fail and leave living microbes behind.
Technical transfer	The formal process of taking a manufacturing process at one location (or on one line) and implementing it at another.
Terminal Sterilisation	Producing the drug product in a clean but not aseptic fashion, then sterilising the final sealed product via heat or irradiation. Usually not appropriate for biological therapeutics.

Trending	Checking to see if testing results, even if still within specifications, are drifting towards being out-of-specification.
Unit operation	As with a process step, this is a distinct piece of the manufacturing process.
User requirements specification	Literally the requirements that the user has for their equipment, building, object, etc.
Validated	Shown to work consistently according to requirements.
Validation	Checking your processes to see if they consistently work as they should.
Worst-Case Load	Used in autoclave validation, this is a full load of difficult-to-sterilise objects.

About the author

Originally from the sunny shores of Australia, CF Harrison currently works in the beer-filled heart of Bavaria. With a PhD in biochemistry, he has worked in drug discovery, as a scientific consultant, and as a regulatory affairs manager for a major international pharmaceutical company.

After realising that his friends from academia had no idea how the pharmaceutical industry actually worked, he decided to help answer their questions once and for all. The Life After Life Science series serves as an introduction into the complex and jargon-filled world of big pharma – perfect for those who are looking at taking their first role in the field.

Those interested in contacting him can drop a line to lifeafterlifescience@harrison-scientific.com.

Other books in the series

Starting out in the pharma industry: Essential knowledge for life scientists

CF Harrison, available in eBook and Paperback on Amazon

Pharmaceutical Regulatory Affairs: An Introduction for Life Scientists

CF Harrison, available in eBook and Paperback on Amazon

Aseptic Production: An Introduction for Life Scientists

CF Harrison, available in eBook and Paperback on Amazon

From test tubes to tonnes: Commercial drug process development for Life Scientists

CF Harrison, available in eBook and Paperback on Amazon

CPSIA information can be obtained
at www.ICGtesting.com
Printed in the USA
LVHW02s0130191217
560223LV00041B/2270/P

9 781547 182596